Asia

and Berries

Asian Fruits and Berries

Growing Them, Eating Them,
Appreciating Their Lore

KATHLEEN LOW

McFarland & Company, Inc., Publishers
Jefferson, North Carolina

Image licenses:

Page 8: https://creativecommons.org/licenses/by-sa/3.0/

Page 12: https://creativecommons.org/licenses/by/2.0/

Pages 16, 19, 39, 40, 49, 53, 64, 72, 104, 124, 129, 134, 143, 155, 174, 179, 194, 195: https://creativecommons.org/licenses/by-sa/3.0/deed.en

Pages 25, 26, 118, 153, 198, 213, 217: https://creativecommons.org/licenses/by/2.0/deed.en

Pages: 31, 50, 199: https://creativecommons.org/licenses/by-sa/4.0/deed.en

Pages 46, 138, 209: https://creativecommons.org/licenses/by/3.0/deed.en

Page 54: https://creativecommons.org/licenses

Pages 108, 210: https://creativecommons.org/licenses/by-sa/2.0/deed.en

Page 182: https://creativecommons.org/licenses/by-sa/2.5/deed.en

LIBRARY OF CONGRESS CATALOGUING-IN-PUBLICATION DATA

Names: Low, Kathleen, author.
Title: Asian fruits and berries : growing them, eating them, appreciating their lore / Kathleen Low.
Description: Jefferson, North Carolina : McFarland & Company, Inc., Publishers, 2019 | Includes bibliographical references and index.
Identifiers: LCCN 2019040784 | ISBN 9781476675954 (paperback) ∞
ISBN 9781476637723 (ebook)
Subjects: LCSH: Fruits—Asia. | Berries—Asia. | Fruits—Asia—Folklore. | Berries—Asia—Folklore.
Classification: LCC SB354.6.A78 L69 2019 | DDC 634.095—dc23
LC record available at https://lccn.loc.gov/2019040784

BRITISH LIBRARY CATALOGUING DATA ARE AVAILABLE

ISBN (print) 978-1-4766-7595-4
ISBN (ebook) 978-1-4766-3772-3

Front cover images © 2019 Shutterstock

Printed in the United States of America

McFarland & Company, Inc., Publishers
Box 611, Jefferson, North Carolina 28640
www.mcfarlandpub.com

To my sister Carrie who always encouraged
me to write about what I love,
and to my parents from whom I inherited
a passion for Asian fruits and gardening.

Table of Contents

The Fruits and Berries

Introduction

Growing up more than a half century ago on a small farm in Northern California, as a child I loved going out into the yard to pick my favorite fresh fruits. I loved eating loquats and Hachiya persimmons. It wasn't until years later that I learned that many of the fruits I loved consuming were considered uncommon, hard-to-find fruit from Asia.

But today in this era of global trade, what once were considered rare exotic fruits can now be found in many supermarkets. Some have even entered the mainstream. But if you're like me, I tend to not spend my hard earned money on an intriguing new fruit I see which I know nothing about. I'd want to know what it may taste like and how to eat it before considering purchasing it.

The goal of this book is to familiarize the reader with Asian fruits and berries. Although there are varieties of fruit commonly enjoyed by Asians, many of them are not native to Asia. This book includes only fruit and berries that originated in Asia.

For each fruit and berry included in this book you will find a description of it including its taste. Since many of the fruit originated centuries ago, and in one case dates back to prehistoric times, a brief history of the plant is included as well as a description of its characteristics. Because of the long history of many of the fruits and berries, you'll also find myths and lore about each of them.

Once you try them, you may fall in love with the taste of many of these fruits. But since it may be hard to regularly find some of the fresh Asian fruits included depending upon where you live, you may want to try your hand at growing the fruit yourself. Based upon the types of "how to grow" questions I receive as a Master Gardener, I decided to include basic information on

1

how to successfully grow the specific fruits or berries regardless of your gardening skill level or knowledge.

Having the proper climate suitable to growing a specific fruit is essential, especially since some of the fruits included in this book are tropical or subtropical, so I've included the USDA Hardiness zones of the fruit plants included. Although the hardiness zones do not reflect the heat or humidity of a zone, they do reflect the minimum cold temperatures the plant can survive in.

Also included is information on how to consume the fruit or berry, and how it is consumed in Asian countries. In many cases the fruits and berries are reportedly canned or made into jams, jellies or other soft preserves in other countries. As a former Master Food Preserver, I know it's important to know that although people may home can and preserve various fruits and berries here and in other countries, there is no guarantee it is safe to do so unless you follow instructions and recipes that are scientifically determined to be safe. To this end, I've included an appendix on food safety that includes basic tips on safely eating fresh fruits and berries and sources of safe canning recipes and instructions.

Some Asian fruits and berries are currently marketed as "super" fruits because they may be high in certain vitamins or antioxidants. But a few of the fruits in this book are potentially harmful to people with certain conditions or who are taking certain medications. Any warnings regarding the consumption of a particular fruit are listed in the Additional Information Section for the fruit or berry.

I want to acknowledge and thank Jennifer Baumbach, Program Coordinator for the University of California Master Gardeners, Solano and Yolo Counties, for lending her expertise and using her personal time to review sections of my book for accuracy. I also want to acknowledge Audrey Dodds and Irene Stone, my supervisors in my first job as a librarian at the California State Library's State Information and Reference Center for not only teaching me excellent reference and research skills and techniques, but for instilling a lifelong love of research which led me to write this fourth book.

Medical Advice Disclaimer

The information on the uses of the fruits in this book is for informational purposes only. It is not a substitute for professional medical advice or treatment for specific medical conditions. You should not use this information to diagnose or treat a health problem or disease without consulting with a qualified healthcare provider. Always seek the advice of your physician or other qualified health provider with any questions you may have regarding consumption of any fruit listed in this book and your medical condition. Never disregard professional medical advice or delay in seeking it because of something you have read in this book. The author does not recommend or endorse consumption of any of the fruit listed in this book for the treatment of any medical condition. Reliance on any information provided in this book is solely at your own risk.

THE FRUITS AND BERRIES

Asian Pears (*Pyrus pyrifolia* [Burm. f.] Nakai and *Pyrus ussuriensis* Maxim)

Asian pears, also know as apple pears, Oriental pears, Chinese pears, Japanese pears, salad pears, and sand pears, tend to be categorized into three different types based upon their appearance. The first type consists of round or flat pears with yellow to green skin. The second consists of round or flat pears with a bronze colored of light russet colored skin. And the third type consists of traditional pear shaped fruit with green or russet colored skin.

Asian pears are juicy, but less juicy than European pears like the Bartlett, and have a crunchy texture like apples. Because of this, they are also known as apple pears, but are not a cross between an apple and a pear. Some cultivars have a slightly grainy texture which is why they are also known as sand pears. Unlike European pears which are picked green, then ripen off tree, Asian pears ripen on the tree like apples and remain crisp.

ASIAN PEARS LORE. Ancient Chinese believed pears were a symbol of immortality because the trees lived for such a long time. The Chinese word for pear, *li*, also means separation, so it's believed that to avoid a breakup, lovers should not divide pears between themselves. And for this reason, pears should not be gifted during the lunar New Year, unless you want them to bring bad luck to the recipient.

During Japan's Edo period (1603–1867), the Northeastern corner of a person's property was considered the Devil's entryway into the property. So Asian pear trees were frequently planted around the corners of property to keep the devil out.

5

Asian pears *(krzys16/pixabay.com)*

TREE HISTORY. Asian pears originated in China and Japan. In China, their cultivation dates back at least 3000 years ago. Pear seeds appeared in Japan as early as the year 200–300 CE. The pears quickly spread to other Asian nations. Chinese immigrants introduced Asian pears into the West Coast of the United States in the 1800s.

TREE CHARACTERISTICS. Asian pear trees tend to be divided into the Japanese varieties, which have round fruits, and the Chinese varieties, which produce pyriform shaped fruit like the European varieties. Chinese cultivars ripen later in the season than the Japanese cultivars. Most Asian pear cultivars available in the United States are Japanese. Although there are hundreds of different Asian pear cultivars, only a couple of dozen are available in this country.

Deciduous trees, they generally grow from 15 to 30 feet tall and 10 to 25 feet wide, although dwarf cultivars are available. The tree features alternate glossy green ovate to oblong leaves measuring slightly over two and a half inches to seven inches. The leaves turn shades of red, yellow and orange in the fall. Its bark is dark brown. The tree bears white blossoms in the spring. Although the trees are self fruitful, cross pollination increases both fruit production and fruit size.

The number of chilling hours required for fruit production varies depending upon the cultivar. Some trees fruit and grow with as few as 550 to 650 chill hours, while others require a minimum of 800 or more chill hours for fruiting. Asian pear trees begin to fruit in two to four years. The trees have a long life span, ranging from 50 to 150 years.

Some popular cultivars are the following: 'Chojuro' (a.k.a. 'Plentiful')—Russet, greenish brown colored skin, large fruit. • 'Hosui' (a.k.a. 'Sweet Water')—Yellowish brown skin, russet, large fruit, juicy pulp. • 'Ichiban' (a.k.a. 'First Pear')—Brown skinned, moderately disease resistant. • 'Kikusui'—(a.k.a. 'Floating Chrysanthemum')—small to medium, smooth yellow green skin. • 'Kosui'—(a.k.a. 'Good Water')—Brown skinned, apple shaped, small to medium sized. • 'Nijisseiki' (a.k.a. 'Twentieth Century')—Very popular cultivar, yellow green skin, sweet but tart flavor, medium sized fruit, unfortunately very susceptible to fire blight. • 'Shinko' (a.k.a. 'New Success'—Brown russet skin, large aromatic pulp, high fire blight resistance. • 'Shinseiki' (a.k.a. 'New Century')—Medium, yellow skinned, sweet mild pulp. • 'Ya Li'—Chinese pear, smooth shiny yellow, tasty pulp.

▶ FAMILY: Rosaceae. GENUS: *Pyrus* L. SPECIES: *Pyrus pyrifolia* (Burm. F.) Nakai and *Pyrus ussuriensis* Maxim.

SELECTING A TREE. When selecting a cultivar, make sure the chilling hours for the specific cultivar meets your climate. Also look for cultivars that are resistant to diseases in your region. Fire blight is a disease which commonly kills Asian pear trees. If you live in an area where fire blight is a problem, look for cultivars that are less susceptible to it.

If you are planning to plant two pear trees for cross pollination, be sure to select two cultivars that blossom at the same time. The second cultivar can be either an Asian cultivar (preferred) or European pear cultivar providing the blooming time is the same.

When selecting a tree from the nursery, check the tree's overall health. Look for active growth, such as new or young leaves. Make sure its leaves are not yellow or discolored, and the tree is free from insects or signs of insect damage. Check the entire tree to make sure it is free from injury, such as broken branches, cuts in the trunk, etc. And if possible, gently lift the tree out of the pot to check the roots. If the roots are overgrown in the pot and are seriously girdled, the tree should be avoided since it will develop a poor root structure.

When selecting a bare root tree from the nursery, check the bud union (i.e., the place where the tree cultivar is grafted to the rootstock). It should be straight. Avoid it if it is bent. Also avoid any plants that have dark colored oozing bark, which could indicate a potential canker caused by bacteria. Also

Asian pear tree blossoms *(Qwert1234/Wikimedia Commons)*

check for any holes that could be caused by insects. And if possible, check to make sure the roots are not damaged.

How to Grow an Asian Pear Tree. Asian pear trees grow best in full sun and in deep, well draining soils. They prefer soils with a pH of 5.9–6.5. Although the trees are self fruitful, planting another Asian pear cultivar is desirable for better fruit production. (Just make sure the trees blossom at the same time.) Trees should preferably be spaced 15 to 20 feet apart.

Once you've selected the best location for the trees, it's time to start digging the holes. Dig a hole for each tree at least twice the diameter of the root ball and as deep as the root ball. Remove the tree from its pot, or plastic bag if it's a bare root tree, and cut away or shorten any girdled roots. The tree should be planted so that the bud union is several (three to four inches) higher than the soil line (to help prevent crown rot). Place the tree in the hole and refill it with the original soil. Water the newly planted tree thorough, and water it regularly to help it become established.

To protect newly planted trees from sunburn, it's advisable to paint the tree stem with a mixture of 50 percent white interior latex paint and 50 percent water. This helps prevent the bark from drying out and cracking, leading it to be more susceptible to insect infestations.

Young trees should be trained. This involves locating the central tree stem, which should be the strongest and straightest (most vertical stem) and pruning away any other stems. Identify the tree's primary scaffold branches (permanent branches that will stay on the tree its entire life) and prune back competing branches. Scaffold branches should be vertically spaced about a foot between branches. Any dead or broken branches should be pruned away.

Asian pear trees require fruit thinning. This involves removing about half the blossoms in each cluster. A second thinning is usually done about two weeks later. The trees are susceptible to fire blight, bacterial cankers, coddling moths, stink bugs, two-spotted spider mites and other plant bugs.

▶ USDA Hardiness Zones: 5–9. Chilling Hours: 350–800. Water Requirements: Average.

How to Consume. Asian pears are most frequently eaten fresh. They can be stored at room temperature for about a week, and in the refrigerator for up to three months. Fresh Asian pears can be added to salads, and can be used in baked goods such as breads, tarts, pies, etc.

Asian pears can also be dried. To dry them in an electric food dehydrator, first thoroughly wash them in clean water, and pat them dry with a paper towel. Next, cut them into slices a quarter- to half-inch wide. The pears should be pretreated to prevent them from darkening. They can be pretreated in ascorbic acid (available from pharmacies in powder form). To pretreat the slices, mix a teaspoon of the powdered acid into two cups of water, and let them soak in the solution for up to five minutes. (Note: the mixture can be used twice before it needs to be replaced.) Remove the slices, and lay them in a single layer on your dehydrator tray.

Note the pears can also be pretreated in lemon juice, or pineapple juice or any other juice high in vitamin C. Again, soak the slices for up to five minutes before removing them to your dehydrator tray. (Again, the juice needs to be replaced after two uses.) Of course, the disadvantage or advantage, however you see it, to pretreating with lemon or other juices is that the pear slices takes on some of the flavor of the juice it was soaked in.

Dehydrate the Asian pears at 140 degrees Fahrenheit. It can take six to 12 hours to dry. To test for dryness, cut a couple of pieces in half. You should not be able to squeeze any moisture out of them, and there should be no visible moisture. And if you fold a piece over, it should not stick to itself. Once the fruit is dry, let it cool for an hour. Then package the fruit preferably in a freezer container or glass jar. Freezer bags can also be used but unfortunately they are not rodent proof. Store the container in a dark, dry, cool place for no more than a year.

They can also be dehydrated in the oven and sun dried (conditions permitting). Both these methods take significantly longer. But you can find instructions of drying fruit using these two methods on the National Center for Home Food Preservation's website at www.nchfp.uga.edu.

Asian pears can also be canned, but must be acidified to prevent botulism. Note different cultivars have varying levels of acidity, but all require additional acidity when canned. For information on how to can Asian pears, you can find instructions on the website of the National Center for Home Food Preservation at www.nchfp.uga.edu.

ADDITIONAL INFORMATION. The tree wood is sometimes used to make furniture.

REFERENCES

"Asian or Oriental Pears." Pennsylvania State University Extension. http://extension.psu.edu/plants/gardening/fphg/pome/variety-selection/asian-or-oriental-pears. Accessed 10/22/2015.

"Asian Pear (*Pyrus pyrifolia*)." Uncommon Fruit, Observations from the Carandale Farm, University of Wisconsin. http://uncommonfruit.cias.wisc.edu/asian-pear/. Accessed 12/21/18.

"Asian Pears." Bulletin no. 5. University of Arizona Cooperative Extension, College of Agriculture and Life Sciences. August 1, 2007. https://cals.arizona.edu/yavapai/publications/yavcobulletins/AsianPears.pdf. Accessed 12/21/18.

Beutel, James A. "Asian Pear." University of Hawaii, College of Tropical Agriculture and Human Resources. https://gms.ctahr.hawaii.edu/gs/handler/getmedia.ashx?moid=3138&dt=3&g=12. Accessed 07/25/18.

Beutel, James A. "Asian Pears." In: *Advances in New Crops. Proceedings of the First National Symposium on New Crops held October 23–26, 1988 in Indianapolis, Indiana*. Portland, OR: Timber Press, 1990. http://fruitsandnuts.ucdavis.edu/dsadditions/_Asian_Pears/. Accessed 10/22/15.

Del Tredici, Peter. "The Sand Pear—*Pyrus pyrifolia*." *Arnoldia: A Continuation of the Bulletin of Popular Information of the Arnold Arboretum, Harvard University*, vol. 67, no. 4, 2010, p. 28–29. http://arnoldia.arboretum.harvard.edu/pdf/articles/2010-67-4-the-sand-pear-pyrus-pyrifolia.pdf. Accessed 12/22/18.

Hill, Lewis, and Leonard Perry. *The Fruit Gardener's Bible*. North Adams, MA: Storey Publishing, 2011.

Ingels, Chuck, et al., editors. *The Home Orchard: Growing Your Own Deciduous Fruit and Nut Trees*. Richmond, CA: University of California, Agriculture and Natural Resources, 2007.

Kendall, P., and J. Sofos. "Preparation: Drying Fruits." Food and Nutrition Series no. 9.309. Colorado State University Cooperative Extension. March 2003. http://extension.colostate.edu/topic-areas/nutrition-food-safety-health/drying-fruits-9-309/. Accessed 04/27/18.

Olcott-Reid, B. "Asian pears." *Horticulture*, vol. 69, no. 7, August/September 1991, p. 68–76.

Parker, David, and Gregg Reighard. "Asian Pear." HGIC 1352. Clemson University, Home Garden and Information Center. September 27, 1999. https://hgic.clemson.edu/factsheet/asian-pear/. Accessed 12/21/18.

Powell, Arlie. "Asian Pear Culture in Alabama." Auburn University Cooperative Extension. http://www.aces.edu/dept/peaches/pearasiancult.html. Accessed 12/21/18.

Preserving Food: Drying Fruits and Vegetables. College of Family and Consumer Sciences, University of Georgia Cooperative Extension Services, July 2000.

"Preserving Foods: Asian Pears." SP 50–64. Oregon State University, Extension Service. February 2015. https://extension.oregonstate.edu/sites/default/files/documents/8836/sp50694 preservingasianpears.pdf. Accessed 12/21/18.

"*Pyrus pyrifolia.*" EcoCrop. Food and Agriculture Organization of the United Nations. http://ecocrop.fao.org/ecocrop/srv/en/cropView?id=9086. Accessed 12/23/18.

"*Pyrus pyrifolia.*" North Carolina State University, Cooperative Extension. https://plants.ces.ncsu.edu/plants/all/pyrus-pyrifolia/. Accessed 12/22/18.

"*Pyrus pyrifolia* (Burm. f.) Nakai." USDA, NRCS. 2018. The PLANTS Database. National Plant Data Team, Greensboro, NC 27401-4901. https://plants.sc.egov.usda.gov/core/profile?symbol=PYPY2. Accessed 12/10/18.

"*Pyrus ussuriensis* Maxim." USDA, NRCS. 2018. The PLANTS Database. National Plant Data Team, Greensboro, NC 27401-4901. https://plants.sc.egov.usda.gov/core/profile?symbol=PYUS2. Accessed12/10/18.

Reich, Lee. "Grow (a ton of) Asian Pears." *Organic Gardening*, vol. 38, no. 5, May/June 1991, p. 57–61.

"Selecting, Preparing and Canning Fruit: Pears, Asian—Halved or Sliced." National Center for Home Food Preservation. https://nchfp.uga.edu/how/can_02/asian_pears.html. Accessed 12/21/18.

"SelecTree: Tree Detail—ASIAN PEAR Pyrus pyrifolia." California Polytechnic University. https://selectree.calpoly.edu/tree-detail/pyrus-pyrifolia. Accessed 12/22/18.

Stremple, Barbara Ferguson, editor. *All About Growing Fruits, Berries and Nuts.* San Ramon, CA: Ortho Books. 1987.

Wetherbee, Kris. "Meet the Asian Pears." National Gardening Association. https://garden.org/learn/articles/view/741/. Accessed 12/22/18.

Asian Persimmons (*Diospyros kaki*)

Also known as Oriental persimmons and Japanese persimmons, the three most commonly grown cultivars in this country are the astringent 'Hachiya' and the non-astringent 'Fuyu' and 'Jiro.' Seedless fruit, you'll only encounter a few small seeds in persimmons that have been cross pollinated. Persimmons are either astringent or non-astringent. Astringent varieties, like the 'Hachiya,' are acidic and bitter unless eaten when really ripe. 'Hachiya' persimmons need to be ripe to the point of being almost mushy soft. 'Hachiya' persimmons are large acorn shaped and orange. Their size ranges from a baseball to small grapefruit. When ripe, their taste is similar to a combination of a sweet apricot and peach.

'Fuyu' and 'Jiro' persimmons, which are oblate shaped, are non-astringent. Smaller in size than 'Hachiya' persimmons, they can be eaten crisp like apples, or somewhat soft when really ripe. Their taste is similar to a cherry infused apple.

Two pollination variant persimmons, the 'Chocolate' persimmon and the 'Cinnamon' persimmon are also gaining popularity. Pollination-variant persimmon trees can produce both astringent and non-astringent tasting persimmons. Tree blossoms that are pollinated produce non-astringent persimmons with seeds. Those blossoms not pollinated produce astringent persimmons with no seeds. Both the 'Chocolate' and 'Cinnamon' cultivars are similar to 'Fuyu' and 'Jiro' persimmons in appearance.

Asian persimmons *(Angelo DeSantis from Berkeley, U.S./Wikimedia Commons)*

PERSIMMON LORE. There's a Korean folk tale about a ferocious tiger with a frightening roar that lived outside a small mountain village. He was hungry so he went down to the village in search of food. He came upon a house with a child inside that wouldn't stop crying. The tiger decided he'd eat the child just to silence him. But as he approached, he saw the child's mother inside who told the child there's a fox and to be quiet or else he'd be eaten by it. The child was not scared by the threat and continued crying. So the mother tried to scare him into quiet by saying now there was a bear outside who was going to eat him. The child was still not frightened. So this time the mother tried scaring the child into silence by saying there was a tiger. The child continued to cry. Seeing the child still wasn't fearful of a great tiger like himself, the tiger decided he would eat the child to end his annoying bawling. Right before the tiger was about to burst into the home and eat the child, the mother told the child to look, there was a persimmon on the table. Upon hearing that, the child became silent. The tiger, seeing that a persimmon was more frightening to the child than a fox, bear, or tiger like himself, thought the persimmon must be a terrible monster. So fearing for his own life, the tiger retreated back to his home.

There's nothing in the folk tale to explain why the child stopped crying at the mention of a persimmon. However, it's often believed that fearing the bitter taste of an unripe astringent persimmon may be the reason why the child stopped crying.

Lots of lore about the medical benefits of persimmons exists. But there's a lack of scientific studies to substantiate the supposed medical benefits. Also lacking are substantial studies on the potential adverse reactions to using persimmons for medical treatment.

In folk medicine, persimmons have been said to help improve blood flow, help prevent strokes, and lower high blood pressure. Eating fresh Asian persimmons (non-astringent, and only ripe astringent varieties) supposedly helps prevent constipation. Cooked astringent persimmons are used to treat diarrhea.

Dried ripe persimmons that have been ground into a powder are used to treat dry coughs, while a powder made from the persimmon calyx is used to cure the hiccups.

TREE HISTORY. Native to China, Marco Polo recorded the trade of Asian Persimmons in the 14th century. Chinese and Japanese immigrants are believed to have first brought the persimmon to America. In 1863 William Saunders from the USDA imported Asian persimmon seeds from which trees were grown and distributed for trial. In 1870 the U.S. Department of Agriculture imported grafted Hachiya and other Asian persimmon trees, some of which were distributed to California, Georgia and other southern states. Persimmon tree trials also began at the University of Florida in the early 1900s.

The USDA Plant Introduction Garden which was located in Chico (CA) began propagating the cultivars previously imported from Asia prior to the 1919 import ban (Federal Horticultural Board Quarantine Order 37). Many of those cultivars were repropagated at UCLA and the University of California Wolfskill Experimental Orchards in Winters (CA). The cultivars from these sites are credited with providing the nucleus of the persimmon industry in California.

The persimmon tree trials planted at the University of Florida at Gainesville were also successful with many cultivars planted. The trees were popular because of their prolific crops in the fall. Reportedly 22,000 trees were being grown commercially at one time.

TREE CHARACTERISTICS. Deciduous trees, non-astringent fruit bearing persimmon trees can grow up to 25 feet tall and equally wide. Astringent fruit producing trees can grow up to 40 feet tall. The branches of astringent fruit bearing trees are somewhat brittle and easily broken. The tree blossoms are small, white, surrounded by a green calyx, and are almost unnoticeable.

The leaves, shaped like a pointed oval, are pale green when young, but turn dark green and somewhat glossy when mature. In the fall, the leaves, especially on the astringent producing trees turn a brilliant yellow, orange and red. All the leaves will fall off long before any fruit left on the tree falls off. The trees go dormant in the winter. Some people consider persimmon trees as landscape trees.

The trees grow best in areas with mild winters and moderate summers. They do not do well in temperatures below 10 degrees. The trees also do not produce a good crop of fruit in extremely high temperatures. Mature trees are drought tolerant, although the decrease in water will decrease the size and quality of the fruit.

▶ FAMILY: Ebenaceae. GENUS: *Diospyros* L. SPECIES: *Diospyros kaki* L.f.

SELECTING A TREE. Be sure to select either an astringent, non-astringent, or pollination variant cultivar tree based upon your personal preference. And also consider the available space for the mature tree.

When selecting a tree from the nursery, check the tree's overall health. Look for active growth, such as new or young leaves. Make sure its leaves are not yellow or discolored, and the tree is free from insects or signs of insect damage. Check the entire tree to make sure it is free from injury, such as broken branches, cuts in the trunk, etc. And if possible, gently lift the tree out of the pot to check the roots. If the roots are overgrown in the pot and are seriously girdled, the tree should be avoided since it will develop a poor root structure.

When selecting a bare root tree from the nursery, check the bud union (i.e., the place where the variety is grated to the rootstock). It should be straight. Avoid it if it is bent. Also avoid any plants that have dark colored oozing bark, which could indicate a potential canker caused by bacteria. Also check for any holes that could be caused by insects. And if possible, check to make sure the roots are not damaged.

GROWING PERSIMMONS. Persimmon trees grow in most soils, except light sandy soils. However they grow best in loamy soil. They tolerate clay soils as long as there is proper drainage.

As you select a location to plant your tree, be sure to select a site in full sun. A preferable site is also shielded from strong winds. Also remember to select a location that will accommodate the height and width of the tree as it matures. Plant the tree in an area protected from high winds that can cause significant fruit scaring. If planting more than one persimmon tree, 'Fuyu' and 'Jiro' trees should be planted at least 10 feet from one another, and 'Hachiya' trees at least 19 feet apart.

If you purchased a bare root tree, remember to soak the roots in a bucket of water for several hours prior to planting. Whether you are planting a bare root or container tree, when planting a tree, be sure to dig the hole as deep as the roots, and twice as wide. The tree should be planted so that the bud union is several (three to four inches) higher than the soil line (to help prevent crown rot.) Place the tree in the hole and refill it with the original soil. Water the newly planted tree thoroughly, and water it regularly to help it become established.

To protect newly planted trees from sunburn, it's advisable to paint the tree stem with a mixture of 50 percent white interior latex paint and 50 percent water. This helps prevent the bark from drying out and cracking, leading it to be more susceptible to insect infestations.

Asian persimmon trees are drought tolerant. But for the best fruit quality fruit, they should be irrigated starting in the spring and tapering off in the fall. The trees should receive an additional 36 to 45 inches of water during this period. The addition of fertilizer is generally not needed. But if you do decide to fertilize, use a light fertilizer. Apply in the spring and early June. If you are using drip irrigation, place the fertilizer under the drip emitters. If not using drip irrigation, spread the fertilizer under the canopy, at least a foot away from the tree trunk, and water it into the soil.

Young trees need to be pruned to develop a strong framework to support a heavy crop. This is especially important for 'Hachiya' trees where heavy crops are known to break branches due to their weight. Mature trees should be pruned annually during the winter while the trees are dormant. The trees are known for alternate bearing—i.e., bearing heavy crops of small persimmons one year, then light crops of large fruit the subsequent year. Thinning the fruit when it is half-inch in size is supposed to reduce the tree's tendency toward alternate bearing.

▶ USDA HARDINESS ZONES: 7–11. CHILLING HOURS: 100–200. WATER REQUIRE-MENTS: Average. Trees should also be irrigated during the spring and summer.

How to Consume. To eat fresh, harvest non-astringent persimmons when they have full color. They can be eaten crisp like an apple, or sliced up and added to salads. 'Hachiya' persimmons should be harvested when they have full color but are still firm. Be careful because they do bruise easily. Store the persimmons until they are soft almost to the point of being mushy. To eat fresh, peel off the skin and scoop out the center with a spoon and eat like pudding. Both types of persimmons are frequently used in baking cookies and breads. They are also used in jams and chutneys. Ripe 'Hachiya' persimmon pulp can also be frozen and stored for later use in baking, or eaten as a frozen treat.

'Hachiya' persimmons are commonly dried. Dehydrating unripe 'Hachiya' persimmons removes the astringency. To dehydrate 'Hachiya' persimmons in an electric food dehydrator, select firm persimmons. Wash the

Air dried 'Hachiya' persimmons; the white coating is the fruit's natural sugar arising from the air drying process *(ProjectManhattan/Wikimedia Commons)*

persimmons then cut away the stem. Peel them. Cut the pulp into slices a quarter- to no more than half-inch. Lay them single layer on your dehydrator tray and set the temperature for 135 to 140 degrees Fahrenheit. It can take up to 10 to 12 hours for them to dry. To test for dryness, cut a couple of pieces in half. You should not be able to squeeze any moisture out of them, and there should be no visible moisture. And if you fold a piece over, it should not stick to itself.

Once the fruit is dry, let it cool for an hour. Then package the fruit preferably in a freezer container or glass jar. Freezer bags can also be used but unfortunately they are not rodent proof. Store the container in a dark, dry, cool place for no more than a year. Non-astringent cultivars can also be dehydrated.

In Asia 'Hachiya' persimmons are frequently air dried whole naturally. It takes about a month to air dry persimmons. To air dry whole persimmons, wash and then peel them, leaving the stems intact. Tie strings to the persimmon stems and hang them in a well ventilated area where air can reach all sides of the fruit. The persimmons are dried when they feel rubbery but still pliable.

Air dried persimmons need to be pasteurized to destroy any insects or their eggs that may be on the fruit. There are different ways to pasteurize them. The first method is to place the persimmons single layer on a baking sheet. Preheat your oven to 160 degrees Fahrenheit and bake for 30 minutes. The other method is to place the persimmons in a sealed freezer bag and freeze it for at least 48 hours at zero degrees Fahrenheit or lower.

The persimmons can also be solar dried, or dried in an oven. For information on how to dry fruit in these methods, visit the National Center for Home Food Preservation's website at www.nchfp.uga.edu. 'Hachiya' persimmons can also be made into regular jam, or freezer jam. But be sure to only use a safe jam recipe—those from a pectin company, canning equipment manufacturing company, university or other source that has scientifically reviewed and or tested the jam recipe for safety. Never simply trust a recipe you found on the internet.

ADDITIONAL INFORMATION. WARNING: Because Asian persimmons could lower your blood pressure, if you are taking high blood pressure medication, you should consult with your doctor before consuming Asian persimmons.

Persimmon tree leaves are also used to make herbal teas.

California grows the most Asian persimmons commercially in the United States, followed by Florida, and then Texas. In 2015 the United States exported 6.9 million pounds of Asian persimmons. Worldwide, the three largest producers of Asian persimmons are China, South Korea, and Japan.

Although there are more than 1000 cultivars of Asian persimmons, only about 100 varieties are grown in the United States. The most common astringent variety grown in America is the Hachiya. Popular non-astringent varieties grown include the Fuyu and Jiro.

Asian persimmons (*Diospyros kaki*) should not be confused with the American persimmon (*Diospyros virginiana* L.), also known as the common persimmon. The earliest written documentation of the common persimmon dates back to the late 1500s.

REFERENCES

Ames, Guy. "Persimmons, Asian and American." National Sustainable Agriculture Information Service, National Center for Appropriate Technology. 2010. www.attra.ncat.org/attra-pub/persimmon.html. Accessed 01/13/18.

Briand, C.H. "The common persimmon (*Diospyros virginiana* L.): The History of an Underutilized Fruit Tree (16th–19th Century." *Huntia*, v. 12, no. 1, 2005, p. 71–89.

Chia, C.L., et al. "Persimmons: General Crop Information." University of Hawaii Extension. n.d. www.extento.hawaii.edu/kbase/crop/crops/i_persim.htm. Accessed 01/18/16.

Das, Soumitri, Laura Shillington, and Tom Hammet. "Persimmon." Fact Sheet #13. Nontimber Forest Products Program, College of Natural Resources, Virginia Tech. 2001. http://asdevelop.org/wp-content/uploads/2016/09/persimmon-VT.pdf. Accessed 03/06/18.

"*Diospyros kaki*." USDA, NRCS. 2018. The PLANTS Database. National Plant Data Team, Greensboro, NC 27401-4901. https://plants.sc.egov.usda.gov/core/profile?symbol=DIKA2. Accessed 08/12/18.

Farrar, Kristen. *Crop Profile for Persimmons in California*. University of California. 1999. https://ucanr.edu/datastoreFiles/391-473.pdf. Accessed 03/18/18.

"Fruit and Nut Review: Oriental Persimmons." Information Sheet 1446. Mississippi State University Extension. 2015. http://fliphtml5.com/kyjt/asps/basic. Accessed 02/22/16.

Ingels, Chuck, et al., editors. *The Home Orchard: Growing Your Own Deciduous Fruit and Nut Trees*. Richmond, CA: University of California, Agriculture and Natural Resources, 2007.

"Japanese Persimmon." WebMD. www.webmd.com/vitamins/ai/ingredientmono-1116/japanese-persimmon. Accessed 01/27/18.

"The Leading Persimmon Producing Countries in the World." World Atlas. https://www.world atlas.com/articles/the-leading-persimmon-producing-countries-in-the-world.html. Accessed 01/13/18.

Marzolo, Gina. "Persimmon." College of Agriculture, California State Polytechnic University, Pomona. 2016. www.agmrc.org/commodities-products/fruits/persimmon/. Accessed 01/13/18.

Morton, Julia F. *Fruits of Warm Climates*. Eugene, OR: Wipf and Stock Publishers. 2003.

Nagy, Steven, and Philip E. Shaw. *Tropical and Subtropical Fruits: Composition, Properties and Uses*. Westport, CT: AVI Publishing. 1980.

Ortho All About Citrus & Subtropical Fruits. Des Moines, IA: Meredith Books. 2008.

Parker, David, and Greg Reighard. "Persimmon." Clemson University Extension, Clemson, SC. 1999. www.clemson.edu/extension/hgic/plants/vegetables/tree_fruits_nuts/hgic1357.html. Accessed 10/17/17.

Parker, M.L. "Growing Oriental Persimmons in North Carolina." Leaflet #377. North Carolina State University Extension. 1993. https://content.ces.ncsu.edu/growing-oriental-persimmons-in-north-carolina. Accessed 03/06/18.

Popenoe, Wilson. *Manual of Tropical and Subtropical Fruits: Excluding the Banana, Coconut, Pineapple, Citrus Fruits, Olive and Fig*. NY: Macmillan Co., 1920.

Ryerson, Knowles. "Culture of the Oriental Persimmon in California." *Bulletin #416 California Agricultural Experiment Station*. January 1927. p. 1–63.

Schaula, Jeff. "Growing Persimmons." Arizona Cooperative Extension, Yavapai County. 2003. https://ag.arizona.edu/yavapai/anr/hort/byg/archive/growingpersimmons.html. Accessed 01/19/16.

Stein, Larry, Monte Nesbitt, and Jim Kamas. "Persimmons." E-611. Texas A&M AgriLife Extension Service. 2015. https://aggie-horticulture.tamu.edu/fruit-nut/files/2015/04/persimmons_2015.pdf. Accessed 01/19/16.

Wallace, R.D. "Commercial Kaki Persimmon Production in Florida." *Proc. Fla. State Hort Soc*. v. 110, p. 161–162. 1997.

Breadfruit (*Artocarpus altilis* [Parkinson] Fosberg)

Breadfruit, also know as 'ulu in Hawaii, is somewhat round to oblong shaped. On average it measures about five to eight inches long and nearly equally wide, about the size of a grapefruit. But depending upon the cultivar, the size can range from that of an apple to a small watermelon. The skin of most cultivars is rough and varying shades of green when immature, turning yellowish-green when ripe. (A lavender colored cultivar exists.) The skin is edible. The pulp is a creamy yellowish-white and somewhat soft when ripe. Most breadfruit cultivars have small seeds in the center, which are edible when cooked. Seedless varieties are available. Breadfruit can be eaten raw, but is most frequently cooked before consumed. When cooked, breadfruit has the consistency and taste of a bland potato. But the taste has also been described as that of an artichoke heart.

Some cultivars are sweeter than others. Some cultivars cook slower than others. And there are also some differences in the taste of the different cultivars.

Breadfruit *(© Hans Hillewaert/Wikimedia Commons)*

BREADFRUIT LORE. Many believe that breadfruit was the cause of the infamous Mutiny on the *Bounty*. Back in 1787 William Bligh took command of the *Bounty*. At the behest of botanist Sir Joseph Banks, Bligh was on a mission to transport hundreds of breadfruit trees from Tahiti to the West Indies. Because of the bountiful fruit produced by mature trees, the young trees were being transported to the West Indies so they could eventually provide an abundant and inexpensive source of food for the slave labor who worked the fields. But before reaching their final destination, the crew, led by Christian Fletcher, mutinied. Some believe that the tyrannical ways of Bligh, forcing the crew to move the hundreds of heavy breadfruit trees below deck every night and dragged back above deck every day, led to the mutiny. Bligh and 18 crew members were forced into an open boat and only given a week's worth of food. The mutineers also dumped all the breadfruit trees overboard. But as you know, Bligh survived by navigating the boat to Timor, in the Indies.

However, in 1793 Captain Bligh was commissioned to make a second trek to Tahiti. This time he successfully conveyed several hundred surviving

Cut breadfruit *(U.S. Pacific Basin Agricultural Research Center)*

breadfruit trees to the West Indies. So Bligh is credited with introducing breadfruit to the Caribbean. Ironically the slaves, for whom the fruit trees were intended to feed, supposedly did not like the taste of the breadfruit and would not eat them.

In Hawaii there's a story about a man named Ulu who died during a great famine. His family buried him by a spring. The next morning his family found a breadfruit tree growing where he had been buried, providing them with enough food to survive through the famine.

In some countries, a tea is made from the leaves of breadfruit trees. The tea is believed to help cure a variety of conditions, ranging from high blood pressure and liver disease to diabetes. In folk medicine, the latex from the tree is used topically to treat skin conditions, sprains and broken bones. Consumption of the diluted tree latex is said to help cure diarrhea and dysentery.

TREE HISTORY. Botanically, the breadfruit genus (*Artocarpus*) is native to Southeast Asia, the Indian continent, and Australasia. It was spread from the South Pacific to the Caribbean islands by travelers. It subsequently spread to Central and South America and many other countries. Currently breadfruit is grown in more than 80 countries.

It's believed to have been brought to Hawaii in the 12th century by Polynesians. It was introduced into Florida in the early 1900s after the USDA imported breadfruit plants from Panama. Hawaii and Florida are the only two states where significant numbers of breadfruit trees are grown.

TREE CHARACTERISTICS. Described as an ultra-tropical tree, there are hundreds of varieties of breadfruit. Growing an average of 40 to 60 feet tall in the United States; depending upon the cultivar, the tree height can range from 25 feet up to 85 feet tall. This evergreen tree possesses a dense canopy. The tree has huge leathery textured ovate leaves split into up to six lobes. The leaves are glossy green on the top side and dull on the bottom side. All parts of the tree contain a sticky milky latex, which has been used as a caulk and glue.

The trees bear both female and male whitish yellow flower clusters. The male flowers appear first, followed by the female flowers. Each cluster has 1500 to 2000 tiny flowers. Trees are propagated by seed, by grafting, and air layering. Trees begin fruiting in two to three years. The trees can bear up to 250 fruit per year. They are known for producing fruit for 50 years up to 100 years.

▶ FAMILY: Moraceae. GENUS: *Artocarpus* J.R. Forst. & G. Forst. SPECIES: *Artocarpus altilis* (Parkinson) Fosberg.

SELECTING A TREE. Make sure you select a cultivar you desire, taking into consideration the fruit taste and your intended method of consumption. Consider the cultivar's tree height and the space you have available for the tree. Although they are ultra tropical trees, some cultivars require more rainfall than others.

Traditionally breadfruit trees were grown from seeds. If you want to grow a traditional breadfruit cultivar, trees grown from seed are fine. But these days most cultivars and newer seedless cultivars are vegetatively propagated. Root cuttings and cut root shoots are used to grow new breadfruit trees. The advantage of propagating a tree from root cuttings or root shoots is that they are the same genetic material as the parent tree. So selecting a tree that was vegetatively propagated will ensure the tree is true to type.

GROWING A BREADFRUIT TREE. Select a sunny location that can accommodate the size of the tree as it matures (although trees can be meticulously pruned as they grow to keep them a smaller size). The tree prefers soil with a pH of 6.1 to 7.4. To plant the tree, dig a hole as deep as the pot the tree came in, and twice as wide. Since breadfruit trees require soil with good drainage (like sandy loam), depending upon your soil, you may need to mix in compost as you refill the hole with the original soil. Plant the tree "high" so that the top of the root ball is a couple inches above the soil line. Water the tree thoroughly after planting. The tree should be watered regularly until it becomes established, in about a year.

If you decide to prune your tree to keep it a minimal size, then your pruning should begin when the tree is two to three years old, and continue annually. The best time to prune is right after the tree has finished fruiting. After pruning, the tree will benefit from application of a light fertilizer. Tree pests include mealy bugs, scales and white flies.

HOW TO GROW FROM SEED. To grow a tree from seed, retrieve seeds from a ripe breadfruit. Clean away any pulp from the seeds and wash them. Plant the seeds in well draining soil and keep them moist. They will germinate in 10 to 14 days. Note that breadfruit seeds lose their viability quickly so they should be planted shortly after removing them from a ripe breadfruit.

▶ USDA HARDINESS ZONE: 11. CHILLING HOURS: 0. WATER REQUIREMENTS: 60–118 inches annually.

HOW TO CONSUME. Since breadfruit has the taste and consistency of potatoes, it is frequently baked, boiled, steamed or fried. Once cooked, it can be used in casseroles, stews, and soups, served as a side dish, made into a salad, or used in other dishes where potatoes are commonly used. Cooked breadfruit can also be cut into strips and eaten with a dip. Ripe raw breadfruit that has become soft can be used to make desserts, like breadfruit pie, or a breadfruit flan. In Hawaii, breadfruit is often used to make *poi*. Breadfruit can also be made into flour. Step by step instructions for making breadfruit flour can be found on the Breadfruit Institute's website at https://ntbg.org/sites/default/files/generaluploads/_Flour_Grinding_Black_White-2.pdf Commercially, breadfruit has been made into chips and pasta.

You can find instructions on how to steam, fry, bake and freeze bread-fruit in the *Brief Breadfruit Basics* fact sheet issued by the Hawaii Homegrown Food Network and the Breadfruit Institute of the National Tropical Botanical Garden. The document can be found online at https://hdoa.hawaii.gov/add/files/2014/05/Brief_Breadfruit_Basics.pdf.

If you're looking for breadfruit recipes for appetizers, salads, entrees or desserts, you can find dozens of them in the *Na Lima Kokua Cookbook*. The cookbook contains recipes from breadfruit cook-offs held in Maui and Hana, and includes recipes for dishes like Coconut Ginger Breadfruit, Breadfruit Shrimp Pasta, Breadfruit Burger, 'Ulu Kim Chi, Marinated Breadfruit Salad with Lilikoi Vinaigrette, Breadfruit Ice Cream, and Breadfruit Macadamia Nut Cake. Best of all, you can download the cookbook for free from the National Tropical Botanic Garden's website at https://ntbg.org/breadfruit/food/recipes.

Breadfruit has also been fermented and pickled. Breadfruit seeds are also edible when cooked. You can bake, boil, or steam cook the seeds and eat them with a light salting. Some people compare their taste to chestnuts.

ADDITIONAL INFORMATION. Dried breadfruit blossom clusters have been burned to repel airborne insects. Scientists from the USDA Agricultural Research Service and Canada's University of British Columbia have studied breadfruit and found that natural chemicals in breadfruit were more effective than DEET in repelling mosquitoes.

The wood from breadfruit trees has been used to construct homes and canoes. It's also been used to make furniture, surfboards, and items like bowls. And of course the wood has been used as firewood. It is said to be termite resistant. The leaves of breadfruit trees are used as fodder. The tree's sticky latex has been used to caulk canoes. Other uses of the latex include trapping insects (like fly paper traps flies), and catching birds.

The planting of breadfruit trees in suitable climates is promoted for food security. The trees are fast growing. The reasons breadfruit trees are being promoted are the huge amount of fruit produced each year, the long fruit bearing life of breadfruit trees, and the nutritional value of the fruit.

REFERENCES

Alexander, Caroline. "Captain Bligh's Cursed Breadfruit." *Smithsonian*, vol. 40, #6, Sept. 2009, p. 56–88.
"*Artocarpus altilis* (Parkinson) Fosberg." USDA, NRCS. 2018. The PLANTS Database. National Plant Data Team, Greensboro, NC 27401-4901. https://plants.sc.egov.usda.gov/core/profile?symbol=ARAL7. Accessed 12/26/18.
Avant, Sandra. "Breadfruit Not So Appetizing to Mosquitos." *Agricultural Research,* November/December 2013, p. 11.
Avegalio, Tusi. "Pacific Gluten Free Breadfruit Flour Regional Industry Development Initiative. IGA Briefing Paper: University of Hawaii Pacific Business Center Program." University of Hawaii Manoa Campus. https://www.doi.gov/sites/doi.gov/files/migrated/oia/igia/upload/38-University-of-Hawaii-Pacific-Business-Center-Program-DOI_OIA-Briefing-Report-Breadfruit-2-19-14.pdf. Accessed 03/24/18.
"Breadfruit *Artocarpus Altilis.*" Dave's Garden. https://davesgarden.,com/guides/pf/go/59617/. Accessed 03/30/18.
Breadfruit Institute of the National Tropical Botanic Garden. https://ntbg.org/breadfruit. Accessed 03/24/18.
Breadfruit Nutritional Value and Versatility. Breadfruit Institute of the National Tropical Botanical Garden, Kalaheo, HI and Hawaii Homegrown Food Network, Holualoa, HI. 2014.
Elevitch, Craig, Diane Ragone, and Ian Cole. *Breadfruit Production Guide: Recommended Practices for Growing, Harvesting and Handling.* 2nd ed. Breadfruit Institute of the National Tropical Botanical Garden, Kalaheo, HI and Hawaii Homegrown Food Network, Holualoa, HI. 2014.
Grant, Amy. "What Is a Breadfruit Tree: Learn About Breadfruit Tree Facts." www.gardeningknowhow.com/edible/fruits/breadfruit/breadfruit-tree-facts.htm. Accessed 03/30/18.
Gross, Lisa. "Productive, Protein-Rich Breadfruit Could Help the World's Hungry Tropics." National Public Radio. August 9, 2016. https://www.npr.org/sections/thesalt/2016/08/09/487094806/productive-protein-rich-breadfruit-could-help-the-worlds-hungry-tropics. Accessed 03/23/18.
Hawkins, James. "Trouble in Paradise." *Norfolk.* April 2017, p. 149.
Little, Elbert L., Jr. and Roger G. Skolmen. *Agriculture Handbook no. 679.* U.S. Forest Service. 1989.
Morton, Julia F. *Fruits of Warm Climates.* Eugene, OR: Wipf and Stock Publishers, 2003.

Orwa, C., et al. "Artocarpus altilis." 2009 Agroforestree Database: A Tree Reference and Selection Guide Version 4.0 www.worldagroforestry.org/sites/treedbs/treedatabases.asp. Accessed 03/26/16.

"Plant a Tree of Life—Grow 'Ulu: Planting Guidelines." Breadfruit Institute of the National Tropical Botanical Garden, Kalaheo, HI and Hawaii Homegrown Food Network, Holualoa, HI. n.d. https://ntbg.org/sites/default/files/generaluploads/How_to_Plant_a_Tree_of_Life2.pdf. Accessed 04/09/18.

Popenoe, Wilson. *Manual of Tropical and Subtropical Fruits: Excluding the Banana, Coconut, Pineapple, Citrus Fruits, Olive and Fig.* NY: Macmillan Co., 1920.

Siler, Julia Flynn. "'Food of the Future' Has One Hitch: It's All But Inedible." *Wall Street Journal.* Nov. 21, 2011.

Sokolov, Raymond. "A Fruit Freely Chosen." *Natural History.* v. 102, no. 9, September 1993, p. 76–80.

Spiney, Laura. "Wonder Food." *New Scientist.* v. 222, no. 2975, June 28, 2004, p. 40–43.

Zerega, Nyree J.C., Diane Ragone, and Timothy J. Motley. "Complex Origins of Breadfruit (*Artocarpus Altilis*, Moraceae): Implications for Human Migrations in Oceania." *American Journal of Botany*, v. 91, no.5, May 2004, p. 760–766.

Breadnut (*Artocarpus camansi*)

The breadnut (*Artocarpus camansi*) should not be confused with the other fruit (*Brosimum alicastrum*) with the same name "breadnut." Visually resembling the breadfruit but smaller in size, at one time the breadnut was considered a seeded variety of the breadfruit until geneticists concluded it is the ancestor of the breadfruit. And despite its common name, the breadnut, also commonly known as a Kamansi, and as a seeded breadfruit, is indeed a fruit and not a nut, although it's commonly grown for its seeds rather than for its pulp.

The breadnut is oval, approximately four to six inches long by two and three-quarters to four and three-quarters inches in diameter. It is smaller than the breadfruit. It has a spiny skin that is dull green to greenish yellow or brown when ripe, covering a whitish flesh. Unlike the breadfruit, the breadnut has little pulp because of the numerous approximate inch long edible cream colored seeds with a brown testa in the fruit. The number of seeds can range anywhere from a dozen to 150 per fruit. But on average the number of seeds per fruit is 56. The many seeds in each fruit can make up 30 to 50 percent of the fruit's total weight. The taste and texture of the seeds is often compared to that of chestnuts.

When ripe, the pulp is said to be sweet. However, if used for cooking, young fruit whose seeds are not yet covered by a hard coat are usually recommended. Young fruit reportedly tastes like a mushroom.

BREADNUT LORE. The breadnut seeds are believed to be an aphrodisiac. The seeds have niacin which is rumored to improve one's sex drive. In folk medicine, breadnut seeds are used to treat arthritis, and the leaves are believed to be effective in treating, diabetes, asthma, and ringworm.

Breadnut *(Colin and Sarah Northway/flickr.com)*

TREE HISTORY. The origin of the breadnut is believed to be Indonesia, New Guinea, and the Philippines. It spread throughout the South Pacific. It was introduced into the Caribbean by French navigators, and then into Jamaica and St. Vincent in 1793, and later to the tropics. In addition to Southeast Asia and the Caribbean, trees can now be found growing in Central and South America, Costa Rica, Western and Central Africa, the Congo, Tahiti, Palau, the Marquesas, and Hawaii.

TREE CHARACTERISTICS. Breadnut trees are tropical evergreen trees reaching 50 or more feet in height. It has a single upright trunk 39 inches in diameter or larger, and a spreading canopy with a diameter about half the height of the tree. Its bark is smooth and gray. The tree features large glossy green alternate leaves measuring 16 to 20 inches wide and 24 to 35 inches long.

Under favorable conditions, the tree is fast growing. It can grow anywhere from one and a half to five feet per year. It develops an extensive shallow tap root system that grows slightly below or along the surface. All parts of the tree possess a white sticky latex.

The tree bears both male and female pale yellow blossom clusters. The male blossom clusters appear a few weeks before the female blossom clusters.

Breadnut on a tree *(Dick Culbert/flickr.com)*

The male blossoms contain viable pollen which fertilizes the female blossoms primarily via bees, insects and the wind.

Although some sources state breadnut trees begin producing crops at eight to ten years of age, a report from the Food and Agriculture Organization of the United Nations states they begin bearing fruit in three years. Mature trees can produce 600 to 800 fruits per year. The trees have a lifespan of 50 or more years. They are propagated by seed. No cultivars of the breadnut tree have been identified.

▶ FAMILY: Moraceae. GENUS: *Artocarpus*. SPECIES: *Artocarpus camansi*.

SELECTING A TREE. Starting out with a healthy tree improves your chances of successfully growing a breadnut tree. When selecting a tree from the nursery, check the tree's overall health. Make sure its leaves are not discolored, and free from insects or signs of insect damage. Check the entire tree to make sure it is free from injury, such as broken branches, cuts in the trunk, etc. If possible, check the tree roots by gently lifting the tree out of the pot. The roots should flare out evenly from the trunk. If the roots are girdled, or kinked around in a circle, the tree is overgrown and may likely develop a poor root structure when planted in the ground.

How to Grow a Breadnut Tree.

How to Grow a Breadnut Tree. Remember that breadnut trees are tropical trees. They grow best in areas with a mean average temperature between 70 and 90 degrees Fahrenheit. Young trees require uniformly moist soil during their first few years. The trees grow best in areas with annual rainfall between 60 and 118 inches distributed throughout the year. Trees will need to be irrigated during dry spells. Extended dry periods create moisture stress in the trees resulting in wilted leaves, leaf drop and stunted growth.

The trees grow in a wide range of soils but prefer well drained soils (loams) with high organic matter. They can grow in neutral to alkaline soils with a pH of 6.1 to 7.4. Mature trees grow best in full sun.

To plant a tree, select a sunny location that can accommodate the size of a mature tree. Dig a hole at least twice the diameter of the pot the tree came in, about as deep as the container. Place the tree in the hole, loosening or cutting away any girdled roots. The tree should be planted so that the top of the root ball is about one to two inches higher than the soil level. Refill the hole with the original soil. Water the newly planted tree thoroughly. Remember these are tropical trees that prefer humidity and require uniformly moist soil.

How to Grow from Seed.

How to Grow from Seed. To grow from seed, retrieve the seeds from ripe breadnut fruit. (Because the seeds germinate easily, some seeds sprout inside ripe fruit that has fallen from the tree.) Be sure to retrieve only shiny, firm, uniform seeds. Wash away any pulp from the seeds. Because the seeds are viable for only a very short period, they should be planted shortly after they are retrieved.

Plant the seeds in a pot or a plastic germination bag filled with a well draining, moisture retentive planting medium. The seeds should be planted twice the depth of the width of the seed. Keep the soil moist, but not wet. The seeds will germinate within seven to 14 days. The germination rate for seeds is usually close to 100 percent.

The seedlings grow rapidly. If you started the seeds in a pot, transplant them into a two gallon pot once their true leaves have hardened. The seedlings are ready to be planted in the field/ground once they reach a height of about two feet. Be sure not to let the seedlings become root bound in either the propagation bag or pot.

► USDA Hardiness Zones: 10–12. Chilling Hours: 0. Water Requirements: 50–150 inches of rainfall annually.

How to Consume. Be aware that the fresh fruit has a very short shelf
life of only a couple of days. The fruit is chilling sensitive so it should not be
refrigerated. The fruit pulp can be sliced and cooked as a vegetable in soups
and stews. The fruit is also frequently added to curry dishes. The seeds can
be roasted, dehydrated or boiled. The taste is similar to chestnuts. To boil
the seeds, wash away any pulp. Boil them in a pot of water for a half hour.
Add salt if desired. Drain, let cool and enjoy. To roast the nuts, clean and
wash away any pulp. Place the nuts single layer on a baking sheet. Bake in
the oven at 350 degrees for 10–20 minutes. Remove. Add salt if desired. Let
cool and consume. The seeds can be ground into a flour "which is more concentrated in energy and proteins than wheat flour."

Additional Information. Breadnut seeds are high in calcium, phosphorus, potassium, iron and niacin. The seeds are a good source of protein.
The tree wood is flexible and easy to carve. It has been used to make statues,
bowls and other objects.

Dried breadnut blossom clusters have been burned to repel airborne
insects. Scientists at the U.S. Agricultural Research Service have studied
breadnut and found that the capric, undecanoic and lauric acids present in
them repel mosquitoes. In a different study, the saponin extracted from breadnut served as an effective insect repellant.

References

Adeleke, R.O., and O.A. Abiodun. "Nutritional Composition of Breadnut Seeds (*Artocarpus
    camansi*). *African Journal of Agricultural Research*, vol. 5, no. 11, June 4, 2010, p. 1273–1276.
"Artocarpus camansi." 2009 Agroforestree Database: A Tree Reference and Selection Guide,
    version 4.0. www.worldagroforestry.org/treedb/AFTPDFS/Artocarpus_camansi.PDF.
    Accessed 09/16/18.
"Artocarpus camansi—Blanco." Plants for a Future. https://pfaf.org/User/Plant.aspx?Latin
    Name=Artocarpus+camansi. Accessed 09/17/18.
Avant, Sandra. "Breadfruit Not So Appetizing to Mosquitos." *Agricultural Research,* Novem-
    ber/December 2013, p. 11.
"Breadnut." Tropical Fruit Farm. www.tropicalfruits.com.my/pdf/Breadnut.pdf. Accessed
    05/27/18.
"Breadnut: *Artocarpus camansi* Blanco." National Tropical Botanical Garden. https://ntbg.
    org/breadfruit/about/species. Accessed 09/17/18.
Hadiyoana, Dianti, et al. "Repellent Activity of Bio-active Agent from Artocarpus camansi
    against Ae. Aegypti." www.academia.edu/6086325/Repellent_Activity_of_Bioactive_
    Agent_from_Artocarpus_camansi_against_Ae._Aegypti. Accessed 09/21/18.
Hari, Akhil, et al. "*Artocarpus*: a Review of its Phytochemistry and Pharmacology." *Journal
    of Pharma Search*, vol. 9, no. 1, 2014, p. 7–12.
Harrynanan, L., and C.K. Sankat. "The Dehydration and Rehydration Characteristics of the
    Seeded Breadfruit or Breadnut Seed." *Canadian Biosystems Engineering*, vol. 50, 2008, p.
    3.37–3.45.

Postharvest Biology and Technology of Tropical and Subtropical Fruits: Acai to Citrus. Amsterdam: Elsevier Science, 2011.

Prescod, Fred. "Comparing Breadfruit, Breadnut and Jackfruit: How Are They Related?" The St. Vincent and Grenadines Association of Toronto Inc. www.vincytoronto.com/flyers/2006/Breadfruit%20BreadnutJackfruit.pdf. Accessed 09/17/18.

Roberts-Nkrumah, Laura B. *Breadfruit and Breadnut Orchard Establishment and Management: A Manual for Commercial Production.* Rome: Food and Agriculture Organization of the United Nations, 2015.

Roberts-Nkrumah, Laura B. *Breadnut and Breadfruit Propagation: A Manual for Commercial Production.* Rome: Food and Agriculture Organization of the United Nations, 2012. www.fao.org/3/a-i3085e.pdf. Accessed 09/17/18.

Salonga, R.B., S. Hisa.k.a., and M. Nose. "Effect of the Hot Water Extract of Artocarpus Camansi Leaves on 2,4,6-Trinitrochlorobenzene (TNCB)-Induced Contact Hypersensitivity in Mice." *Biological and Pharmaceutical Bulletin*, vol. 37, no. 3, 2014, p. 493–7.

Species Profiles for Pacific Island Agroforestry. Agroforestry Net. Inc. April 2006. http://pacific schoolserver.org/content/_public/Local%20Topics/Pacific%20Islands/Agriculture%20for%20Islands/Specialty%20crops/Traditional%20trees%20of%20Pacific%20Islands.pdf. Accessed 09/18/18.

Valencia, R., and S. Bismark. "Benefits of Breadnut." *Philippines Daily Inquirer*, August 13, 2016.

Buddha's Hand (*Citrus medica* var. *sarcodactylis*)

The Buddha's Hand, also known as a fingered citron, is a citrus fruit with little to no pulp. It's easily recognized by its long finger-like segments. The fruit resembles a human hand and generally measures from six to 12 inches long. The outer rind turns from green to yellow to a hint of orange yellow when mature. The fruit features a white pith and generally no pulp, and thus no juice. Seeds are rarely found in the fruit. The outer rind has an intense citrus fragrance and a strong lemony taste.

There are generally two types of varieties, the open hand variety and the closed hand. In the open hand, the citron's fingers are extended outwards like an open hand. In the other variety the citron's segments are extended somewhat straight and close together. Because of its lack of pulp, the Buddha's Hand is not eaten raw.

BUDDHA'S HAND LORE. In China, the Buddha's Hand symbolizes happiness, longevity and good fortune. It's often displayed in homes and at temple alters. It's also gifted during New Year's as a symbol of good luck for the New Year.

When displayed on altars, some people prefer the "closed" hand fruit, believing it symbolizes prayers. Whereas people who are giving the fruit as gifts tend to prefer the "open" hand fruit believing it represents the giving of good fortune and luck.

In her book *The Land Where Lemons Grow*, Helena Atlee writes about the myth of Harmonillus and the origin of the Buddha's Hand. Harmonillus

Buddha's hand *(Kaldari/Wikimedia Commons)*

was a young man with a beautiful singing voice. But his singing angered a witch to the point that she decided to transform him into a citron tree. His legs became tree roots, and his hands transformed into fingered citrons.

TREE HISTORY. The exact origin of the citron is unknown. However seeds were found in a Mesopotamian dig site dating back to 4000 BCE. Historians tend to disagree as to whether the seeds were brought from the Mediterranean to Asia, or from Asia to the Mediterranean. But it's generally believed the Buddha's Hand originated in Asia, possibly in China or Northwest India. They have been growing in both countries since at least the fourth century CE.

The Buddha's Hand was known to ancient Egyptians, having appeared in Egyptian wall drawings. It made its way to Rome in 300 CE. And it was mentioned in Jewish sources in the 13th century.

Buddha's hand was introduced into the United States from Japan in the late 19th century. It's believed Asian immigrants who came to mine for gold during California's Gold Rush brought the fruit with them to California. Commercial growth of Buddha's Hand is very small in the United States and limited to California, Florida and Hawaii.

Immature Buddha's Hand on tree *(Saga70/Wikimedia Commons)*

TREE CHARACTERISTICS. The Buddha's Hand tree is a small citrus tree or shrub reaching heights of 6 to 15 feet with a spread potentially up to 12 feet. Most trees, however, tend to be six to eight feet tall. An evergreen tree, it bears oblong green leaves measuring four to six inches long. New leaves are initially purple, turning green as they mature. The trees branches have long thorns. A self fruitful tree, it bears small clusters of white blossoms with pale purple highlights. The blossoms are pollinated by bees.

Since seeds are rarely found in Buddha's Hands, the tree is generally propagated by cuttings, or by budding. (Budding is a process where a single bud is grafted onto a suitable rootstock, in this case orange, lemon or grapefruit rootstock.) Over a dozen varieties are available in Asia. A very limited variety of cultivars are available here.

▶ FAMILY: Rutaceae. GENUS: *Citrus.* SPECIES: *Citrus medica.* VARIETY: *Sarcodactylis.*

SELECTING A TREE. When selecting a tree from the nursery, check the tree's overall health. Look for active growth, such as new or young leaves. Make sure its leaves are not yellow or discolored, and are free from insects or signs of insect damage. Check the entire tree to make sure it is free from injury, such as broken branches, cuts in the trunk, etc. Check to make sure the tree trunk is not sunburned. And if possible, gently lift the tree out of the pot to check the roots. Overgrown trees that have seriously girdled roots will likely develop a poor root structure so they should be avoided.

How to Grow a Buddha's Hand Tree. The Buddha's Hand tree grows in a wide variety of soils, but grows best in well draining loam soils. The tree prefers full sun to partial shade. The best time to plant a Buddha's Hand tree is in the spring when there's no longer any danger of frost. Planting in the spring also allows the tree to become established before the winter.

To plant a tree, select a sunny location or one with partial shade that can accommodate the size of a mature tree. Dig a hole at least twice the diameter of the pot the tree came in and about as deep. Place the tree in the hole, loosening or cutting away any girdled roots. The tree should be planted so that the top of the root ball is about one to two inches higher than the soil level. Refill the hole with the original soil. Water the newly planted tree thoroughly. The tree should be regularly watered the first two years.

Young citrus trees are sensitive to frost. If temperatures are expected to dip to 29 degrees Fahrenheit or lower, you need to protect the tree from frost. You should wrap the tree trunk and branches in insulation material like cardboard, fiberglass or frost protection covers. Make sure the soil is moist and not dry, since damp soil retains and radiates more heat than dry soil. Bare soil also radiates more heat than soil covered with mulch or other ground covers.

If your tree does sustain frost damage, don't prune away any dead branches until the spring. This allows time for the tree to recover in warmer weather. It also allows you to better identify the damaged branches to be removed. Young citrus trees benefit from fertilization. If using a citrus fertilizer, follow the instructions on the label. Buddha's Hand trees only need pruning to remove dead branches. They can also be pruned if you want to keep them a certain height.

▶ USDA Hardiness Zones: 9–11. Chilling Hours: 0. Water Requirements: Average.

How to Consume. As with any citrus, the Buddha's Hand should be thoroughly washed before preparing the peel for consumption. Some sellers have reported cleaning citrus fruit with a cloth sprayed with cooking oil before placing their citrus out on display for sale. For food safety, citrus peel should be thoroughly washed with water before consuming. Buddha's Hand zest can be substituted for lemon zest. The peel can be slivered and added to salads, or sprinkled onto fish for flavoring.

Buddha's Hand is frequently candied. There are many methods of making candied citrus peel, some using corn syrup and other sweeteners. If you've never tried making candied citrus peel, following is one common simple method. To candy the fruit, wash it and pat it dry with a clean paper towel. Slice the "fingers" lengthwise into quarter-inch strips. In a pot, boil equal parts of sugar and water, for example two cups water and two cups sugar, stirring until the sugar is dissolved. Add in the Buddha's Hand strips and simmer for 20 to 45 minutes, or until the strips turn translucent. Remove the strips and place them on a wire rack to cool. Once they've cooled, lightly sprinkle the strips with granulated sugar so they are evenly coated.

Another method to candy the fruit is to boil the strips in boiling water for about ten minutes until the peel is tender, remove the peels then place the strips on a wire rack for about 15 minutes. Then add the strips to a boiling solution of equal parts water and sugar for 15 to 30 minutes, remove, cool, and coat with sugar.

Another popular use of the Buddha's Hand is to infuse it into alcohol. Vodka is frequently used, and gin to a lesser extent. Infusing it into alcohol is simple. Wash and slice the Buddha's Hand. Place the fruit into an airtight jar, like a canning jar. Pour the vodka over the fruit. Seal the jar and shake it a few times. On a daily basis shake the jar. Let the fruit infuse, i.e., let the jar sit for five days to two weeks, depending upon how strong you want the flavor to be. When it's ready, strain the fruit from the alcohol and enjoy.

As a citrus fruit, Buddha's Hand is often mixed with other citrus and used in marmalade recipes. It's also pureed and used in baked goods, like cakes, cookies.

ADDITIONAL INFORMATION. Because of their strong fragrance, Buddha's Hands are often placed in homes as air fresheners. In Asia, they've been used to scent clothes and laundry.

REFERENCES

Attlee, Helena. *The Land Where Lemons Grow: The Story of Italy and Its Citrus Fruit*. London: Penguin UK, 2014.
"Buddha's Hand." Specialty Produce. www.specialtyproduce.com/produce/Buddhas_Hand_2036.php. Accessed 07/25/18.
"Buddha's Hand Citron." Citrus Variety Collection, University of California, Riverside. www.citrusvariety.ucr.edu/citrus/buddha.html. Accessed 02/26/17.
"Buddha's Hand (Citron)." Urban Harvest, Inc. http://urbanharvest.org/documents/118591/2583341/Buddha%27s+Hand+%28Citron%29%202014.pdf/41627b0b-6769-45fc-bc6d-0e089cc88141. Accessed 07/16/18.
"Buddha's Hand Is the Fruit You Should Try." *Herald-Times* (Bloomington, IN). December 24, 2014.
"*Citrus Medica* var. *Sarcodactylis*." Missouri Botanical Garden. www.missouribotanical garden.org/PlantFinder/PlantFinderDetails.aspx?taxonid=291425&isprofile=0&. Accessed 07/27/18.
David, Cynthia. "Fresh Bites: Buddha's Hand." *Toronto Star*. December 12, 2013.

"Fingered Citron, Buddha's Hand." Dave's Garden. http://davesgarden.com/guides/pf/go/
54295/. Accessed 11/04/15.
Geisel, Pamela M., and Carolyn L. Unruh. *Frost Protection for Citrus and Other Subtropicals.*
ANR Publication 8100. University of California, Division of Agriculture and Natural
Resources, 2003.
Karp, David. "Introducing … Buddha's Hand." *Los Angeles Times.* Food Section. October 28,
2009.
Kethum, Dan. "Propagating Buddha's Hand Citron." http://homeguides.sfgate.com/
propagating-buddhas-hand-citron-29454.html. Accessed 07/25/18.
Marks, Michael. "What's in Season: Buddha's Hand." *San Jose Mercury News.* September 6,
2013. www.mercurynews.com/2013/09/06/whats-in-season-buddhas-hand/. Accessed
07/25/18.
Morton, Julia F. *Fruits of Warm Climates.* Eugene, OR: Wipf and Stock Publishers, 2003.
Ortho All About Citrus and Subtropical Fruits. Des Moines, IA: Meredith Books, 2008.
Shemesh, Abraham Ofir. "The Fingered Citron and the Dibdib Citron for the Ritual of the
Four Species in Medieval and Modern Literature." *The Torah u-Madda Journal,* vol. 16,
2012/2013, p. 173–185.
Smith, K. Annabelle. "What the Heck Do I Do with a Buddha's Hand?" *Smithsonian Magazine.*
February 24, 2014. www.smithsonianmag.com/arts-culture/what-heck-do-i-do-buddhas-
hand-citron-180949871/?no-ist. Accessed 07/19/18.
Spurrier, Jeff. "Buddha's Hand Citron: Like Lemons, But Zestier." *Los Angeles Times.* L.A. at
Home Blog. May 22, 2012. http://latimesblogs.latimes.com/home_blog/2012/05/buddhas-
hand-citron.html. Accessed 07/25/18.

Calamondin (×*Citrofortunella microcarpa*)

Also known as calamansi, calamonse, kalamansi, calamonding, Philippine
lime, and a miniature Chinese orange, this small citrus fruit measures from
one inch to one and three-quarter inches in diameter and resembles a small
tangerine. It features an orange rind that can range from a yellow orange to
an orange red depending upon the cultivar. The rind is thin. The pulp is orange,
juicy, and is acidic and sour. Its flavor is similar to a lemon or lime.

CALAMONDIN LORE. Although there's a lack of substantial research to
back these claims, calamondins have been said to be a cure for everything
from skin conditions to dandruff and body odor.

TREE HISTORY. The calamondin is a natural hybrid resulting from a cross
of a kumquat and what many suspect to be a mandarin orange. The calamondin originated in China and spread to the Philippines and Indonesia long
ago. From there it spread to Malaysia, India, and southern Asia, Central America, and the West Indies. It was introduced into the United States in 1899.

TREE CHARACTERISTICS. Although calamondin trees can grow up to
25 feet tall, most cultivars sold in the United States are dwarf varieties growing
only up to about six feet. Like other citrus trees, the calamondin is an evergreen
tree. Its leaves are broad, glossy green and usually measure up to three inches

Calamondin *(Michael Matera/pixabay.com)*

long. Its branches are almost free from thorns. The trees bear sweet fragrant blossoms. They're white, about an inch wide, and appear in groups of two or three. The trees are self pollinating.

The trees have a deep tap root, and are hardy down to 20 degrees Fahrenheit. They are often grown in containers as an ornamental tree. Calamondin trees can be grown as bonsai trees. They are also frequently grown as hedges. The trees can be propagated by seed and also vegetatively (from softwood or semi-hardwood cuttings). Trees usually begin bearing fruit starting in their second year.

▶ FAMILY: Rutaceae. GENUS: × *Citrofortunella* J.W. Ingram & H.E. Moore [Citrus × Fortunella]. SPECIES: × *Citrofortunella microcarpa* (Bunge) Wijnands [*Citrus reticulata* × *Fortunella japonica*].

SELECTING A TREE. When selecting a tree from the nursery, check the tree's overall health. Look for active growth, such as new or young leaves. Make sure its leaves are not yellow or discolored, and free from insects or signs of insect damage. Check the entire tree to make sure it is free from injury, such as broken branches, cuts in the trunk, etc. Also visually inspect the pot soil to make sure there is no fungi growing from the soil, indicating uneven or over watering that can affect the overall health of the trees.

Before purchasing a tree, decide whether you want and have the space for a regular tree, or a dwarf cultivar, like the 'Dwarf Variegated' (with yellow and green marbles leaves), or 'Espallier PC 18' which grows to a height of under four feet.

How to Grow a Calamondin Tree. The tree tolerates most soils, but grows best in well draining loamy soils with a pH of 6.0–7.5. The tree also requires full sun or partial shade. It grows best in temperatures between 70 to 90 degrees Fahrenheit.

The best time to plant a tree is in the spring when all danger of frost has passed. Select a sunny location or one with partial shade that can accommodate the size of a mature tree. Dig a hole at least twice the diameter of the pot the tree came in, about as deep as the container. Place the tree in the hole, loosening or cutting away any girdled roots. The tree should be planted so that the root ball is about one to two inches higher than the soil level/grade. Refill the hole with the original soil. Water the newly planted tree thoroughly. As a young citrus tree, it would benefit from a citrus fertilizer three times a year (winter, late spring, and late summer). Follow the directions on the fertilizer label.

Young calamondin trees are very sensitive to frost. If temperatures are expected to dip to 29 degrees Fahrenheit or lower, you need to protect the tree from frost. You should wrap the tree trunk and branches in insulation material like cardboard, fiberglass or frost protection covers. Make sure the soil is moist and not dry, since damp soil retains and radiates more heat than dry soil. Bare soil also radiates more heat than soil covered with mulch or other ground covers. If your tree does sustain frost damage, don't prune away any dead branches until the spring. This allows time for the tree to recover in warmer weather. It also allows you to better identify the damaged branches to be removed.

How to Grow from Seed. To grow a tree from seed, remove the seeds from the calamondin. Wash any pulp away from the seeds. Soak the seeds in clean water overnight. Place the seeds about half-inch deep in a sterile moist planting/potting medium. Cover the pot with plastic wrap and place it in a sunny location indoors. The seeds germinate best in temperatures around 60 degrees Fahrenheit. The seeds should germinate in two to three weeks. Once the seeds sprout, transplant them into a larger pot and move them to a sheltered location outdoors.

▶ USDA Hardiness Zones: 9–11. Chilling Hours: 0. Water Requirements: Average.

How to Consume. Because calamondins taste like lemons and limes, they can easily be substituted for them. Calamondins can be juiced for a refreshing beverage. They can be sliced and added to iced tea or water for some flavoring, or you can freeze the juice into an ice cube and add the ice cube to your tea or other beverage. The juice is also mixed with liqueurs to make flavored alcoholic beverages. It's frequently mixed with other citrus fruits to make marmalade. You can use it to season fish, in place of lemons. You can also use the fruit in baked goods. Several recipes for calamondin cake exist. Calamondins are also used in a variety of desserts ranging from pudding to sorbet.

The citrus peel can also be dehydrated. To dehydrate the peel, first thoroughly wash the calomondin in water, rubbing its surface, and rinsing it under running water. Then pat it dry with a clean paper towel. Next, remove the peel (one eighth to one sixteenth of a inch). Do not include the pith. Place the peel pieces single layer cut side up on your electric dehydrator tray. Set the temperature at 130 to 135 degrees Fahrenheit. Although it could potentially take up to 12 hours to dry, you should dehydrate the peels only until they are dry, careful not to over dry them. To test them to see if they are dry, remove a couple of pieces from the dehydrator and let them cool off. Then slice them in half. You should not see any moisture in them, nor should you be able to squeeze any moisture out of them. The peel should still be pliable, but not sticky, which means if you fold it in half it should not crack, nor should it stick to itself.

When the peel has been dried, remove it from the dehydrator and let it cool for an hour. Then place the dried peel in an airtight container, like a canning jar or plastic freezer bag, marking the date on the container, and store it in a dark cool (60 degrees Fahrenheit) place. Use the peel within a year, and be sure to discard any moldy or spoiled dried peel.

The calamondin peel can be candied. There are several methods of candying the peel. Following is one method: Wash it thoroughly and pat it dry. Slice the peel into quarter-inch strips. In a pot, boil equal parts of sugar and water, for example two cups water and two cups sugar, stirring until the sugar is dissolved. Add in the peel strips and simmer for 20 to 45 minutes, or until the strips turn translucent. Remove the strips and place them on a wire rack to cool. Once they've cooled, lightly sprinkle the strips with granulated sugar so they are evenly coated.

Another method to candy the fruit is to boil the strips in boiling water for about 10 minutes until the peel is tender, then remove and place the strips on a wire rack for about 15 minutes. Then add the strips to a boiling solution of equal parts water and sugar for 15 to 30 minutes, remove, cool, and coat with sugar.

ADDITIONAL INFORMATION. In the Philippines calamondin juice is used to bleach away ink stains from cloth.

REFERENCES

"Calamondin." IDtools. http://idtools.org/id/citrus/citrusid/factsheet.php?name=Calamondin. Accessed 07/21/18.

"Calamondin." Citrus Variety Collection, University of California, Riverside. www.citrus variety.ucr.edu/citrus/calamondin.html. Accessed 07/17/18.

"Calamondin." Stark Brothers Nursery. www.starkbros.com/products/fruit-trees/citrus-trees/calamondin-orange. Accessed 07/21/18.

"Calamondin Orange (X Citrofortunella microcarpa)." Florida State University. www.facilities.fsu.edu/sights/?tag=calamondin. Accessed 07/17/18.

"Calamondin—The Most Versatile Citrus." Texas A & M University. http://aggie-horticulture. tamu.edu/patiocitrus/Calamondin.html. Accessed 08/24/17.

"×Citrofortunella microcarpa (Bunge) Wijnands [Citrus reticulata × Fortunella japonica]." USDA, NRCS. 2018. The PLANTS Database. National Plant Data Team, Greensboro, NC 27401-4901. https://plants.usda.gov/core/profile?symbol=CIMI4. Accessed 07/17/18.

"Citrus for Hawai'i's Yards and Gardens." F&N-14. College of Tropical Agriculture and Human Resources, University of Hawaii at Manoa. June 2008. www.ctahr.hawaii.edu/oc/free pubs/pdf/F_N-14.pdf. Accessed 07/20/18.

Geisel, Pamela M., and Carolyn L. Unruh. *Frost Protection for Citrus and Other Subtropicals*. ANR Publication 8100. University of California, Division of Agriculture and Natural Resources, 2003.

McManiman, Michael, and Trish Wesley. "The Portable, Pottable Orchard." *Horticulture*, vol. 97, no. 5, June 2000, p. 18.

Morton, Julia F. *Fruits of Warm Climates*. Eugene, OR: Wipf and Stock Publishers, 2003.

Nagy, Steven, editor, et al. *Fruits of Tropical and Subtropical Origin: Composition, Properties and Uses*. Lake Alfred, FL: Florida Science Source Inc., 1990.

Ortho All About Citrus and Subtropical Fruits. Des Moines, IA: Meredith Books, 2008.

Siebert, Toni, and Tracy L. Kahn. "Tried and True of Something New? Selected Citrus Varieties for the Home Gardener." University of California, Agriculture and Natural Resources. Publication 8472. November 2011. https://anrcatalog.ucanr.edu/pdf/8472.pdf. Accessed 07/23/18.

Sloane, Lauren. "Calamondin Orange." *Organic Gardening*, vol. 53, no. 1, December 2005/January 2006, p. 13.

Snart, Jennifer E., Mary Lu Arpaia and Linda J. Harris. *Oranges: Safe Methods to Store, Preserve and Enjoy*. Publication 8199. University of California, Division of Agriculture and Natural Resources, 2006. http://homeorchard.ucdavis.edu/8199.pdf. Accessed 01/03/18.

Williams-Dennis, Leslie. "You Eat It Whole: Calamondin Fruit? Peel and All? Makes for Tasty, Unusual Treat." *The Brownsville Herald* (Texas). March 7, 2007.

Carambola (*Averrhoa carambola*)

Commonly known as a star fruit, carambola are oblong fruits measuring three to six inches long and generally have five ribs or corners that run up and down the fruit so that when it is sliced horizontally, the slices look like stars. The skin is thin and waxy and turns pale green with yellow, or yellow, when ripe, and is edible. The fruit pulp is translucent yellow, crispy and juicy. Each fruit can have up to a dozen seeds ranging in size from quarter-inch to half inch each.

There are two varieties of the fruit. One is known as the sweet type, and the other as the tart type. The sweet types tend to be larger with wider ribs than the tart or sour types.

Carambola *(Mailamal/Wikimedia Commons)*

The taste of a carambola is often described as being similar to a plum, with the ripe sweet varieties tasting like a sweet plum, and the tart varieties tasting like a somewhat sour unripe plum. It's also been written that the "flavor hints of plum, with floral and tangy citrus notes." Yet another description states that "most taste like a cross between a grape and an apple."

CARAMBOLA LORE. In Vietnam there's a popular legend about the carambola told to children as a warning about the perils of greed. There are many versions of the myth, but basically the legend starts with a father of two sons. In one version of the legend, the father is a king, yet in other versions he's simply a very wealthy man. The man's eldest son was very greedy and selfish, while the younger son was very kind and generous. When the father died, his oldest son inherited his huge fortune, and his younger son received only a hut with a carambola tree growing nearby. The older son lived in luxury while his brother lived in poverty.

The younger son cared for the carambola tree, in hopes that when it started to bear fruit he could sell the fruit for money. Soon the tree began to bear fruit. But before the younger son could harvest the star fruit to sell, a bird came and ate all the fruit. Seeing all the fruit had been eaten, the younger

Sliced carambola *(SMasters/Wikimedia Commons)*

son lamented the loss of his crop. Hearing the son's laments, the bird told him he would pay him for the fruit he ate with riches. All he needed to do was show up the next day with a three foot bag. With nothing to lose, the son obeyed. The next day, the bird returned and magically transported him to a deserted island covered with riches. You can fill your bag with the gems and gold, the bird told him. He filled up his bag, and the bird flew him home.

Because of his new found wealth, the younger son was now able to live comfortably. Being good hearted, he often helped out others who needed help.

His older brother, hearing of his brother's new found wealth, asked him how he came into his fortune. When his brother told him the story, the older brother offered to trade his home for his brother's hut and the carambola tree. Not being a fool, the younger brother accepted the offer.

When the carambola fruit ripened on the tree the following year, the bird once again visited. The older brother offered the bird a bag of carambola fruit in exchange for a trip to the island. The bird agreed, telling the older brother to bring a three foot bag with him tomorrow. The next day bird the showed up and carried the older brother to the island of riches. But the older brother failed to follow the bird's instructions. Instead he brought a bag twice the size as instructed so he could greedily take more treasure from the island. As the bird was flying the brother home, he was too heavy with his large bag of treasure. The bird couldn't hold onto him and had to drop him into the sea. Clutching his bag of riches, he ended up drowning.

TREE HISTORY. The tree is believed to have originated in Sri Lanka and the Maluk Islands (an East Indonesian group of islands). It's widely grown in Southeast Asia and the Indian continent. It was introduced into Australia and the United States over 100 years ago.

TREE CHARACTERISTICS. Subtropical and tropical trees, carambola trees are evergreen trees in the proper climate. They can grow up to 33 feet tall, and spreading up to 25 feet in diameter. They feature a rounded canopy. Its leaves are green, oval, compound, and measure six to 12 inches long and one and a half to three and a half inches wide. The leaves have a smooth upper surface, with really fine hairs on the underside. The tree bears small pink to purplish pink blossoms on purplish red stems. The trees are self fertile. The trees stop growing at temperatures below 65 degrees Fahrenheit. And freezing temperatures can severely damage and even kill young trees. In cooler climates, the trees become deciduous.

The trees are interesting because once they begin to produce blossoms, they can be induced to do so more than once per year. Pruning lateral shoots from small limbs, and pruning away "whips" induces flowering. Carambola trees are propagated by seed. Trees propagated by seed are generally not true to type. They are also vegetatively propagated (by veneer and cleft grafting). Propagation by air layering is usually not done because of the low success rate and poor root development from this process. Trees begin bearing fruit after two to three years. Mature trees can produce up to 350 pounds of fruit per year.

The trees can be grown in containers as an ornamental tree. This allows the tree to be grown indoors if you live in a cooler climate zone than USDA zone 9.

▶ FAMILY: Oxalidaceae. GENUS: *Averrhoa* Adans. SPECIES: *Averrhoa carambola* L.

SELECTING A TREE. There are numerous cultivars to select from. Your first consideration is whether you want to grow a tree that bears sweet carambolas, or tart carambolas. If you live in Florida, the Institute of Food and Agricultural Sciences at the University of Florida recommends the following six cultivars for the home gardener: 'Arkin,' 'Fwang Tung,' 'Kajang,' 'Kary,' 'Lara,' and 'Sri Kembangan.' Note that the 'Kary' cultivar was developed by the University of Hawaii for commercial growers in that state. At present there appears to be no specific cultivars recommended for California growers.

When selecting a tree from the nursery, check the tree's overall health. Make sure its leaves are not discolored, and that that the tree is free from insects or signs of insect damage. Check the entire tree to make sure it is free from injury, such as broken branches, cuts in the trunk, etc. Check to make sure the tree trunk is not sunburned. Also check to make sure the roots are not seriously girdled, since the roots may not grow properly when planted in the ground.

How to Grow a Carambola Tree. The trees tolerate most soils except high alkaline soils. But they grow best, and produce the most fruit in well draining, rich loam soils. They prefer a moderately acidic soil with a pH between 5.5 and 6.5. The tree should be shielded from constant winds. They are intolerant of winds. Constant wind exposure can cause leaf damage (from leaf drop to distorted leaves) and reduced fruiting, and can stunt the tree's growth.

To plant a tree, select a sunny location that can accommodate the size of a mature carambola tree. Dig a hole at least twice the diameter of the pot the tree came in, and at least as deep as the container. Place the tree in the hole. The tree should be planted so that the root ball is about one to two inches higher than the soil level/grade. Refill the hole with the original soil. Water thoroughly. The trees grow best and fruit best in well draining moist soils. Regular watering is recommended from blossoming to fruit set.

Young trees benefit from a quarter- to half-pound of N-P-K every 30 to 60 days. Mature trees should receive N-P-K fertilizer four to six times per year. During the first two years the tree should be pruned to increase branching. Older trees can be pruned to control the tree height if desired.

How to Grow from Seed. If you intend to grow a tree from seeds taken from a fresh carambola, select only those seeds that are fully developed and plump. Wash the seeds and plant them in damp peat moss or a well draining planting medium. Keep the planting medium moist until germination. Because the seeds are only viable for a few days, they should be planted shortly after removing them from the fruit. In ideal temperatures, the seeds will germinate in about a week, but can take up to three weeks in cooler temperatures.

▶ USDA Hardiness Zones: 10–11. Chilling Hours: 0. Water Requirements: Minimum 70 inches rainfall per year.

How to Consume. Unripe immature carambolas are astringent. But they are often cooked and used as a vegetable. They're also frequently added to curry and stir fry dishes.

Carambolas do not ripen once harvested. Sweet ripe carambolas can be eaten fresh. Simply wash them with water, dry them with a clean paper towel, slice them and enjoy. Sliced ripe carambolas make a visually stunning addition to salads and cocktails. You can add them to other fruit and blend them into a fruit smoothie. They are also often pureed and added to baked sweets, puddings and ice cream. Fresh carambolas can be stored in an airtight container for up to 21 days in the refrigerator. You can also dry them in an electric food dehydrator, or in the oven.

ADDITIONAL INFORMATION. WARNING: The National Kidney Foundation warns that anyone with kidney disease should avoid eating carambola fruit. The fruit contains neurotoxins that people with kidney disease are unable to process and pass from their bodies.

Carambolas are high in vitamin A, C, potassium and fiber. They also contain polyphenols, antioxidants that fight cardiovascular inflammation. Plus, they are low in calories. Carambola fruit and leaves have been used in folk medicine in other countries to treat a wide range of conditions ranging from fever to worm infestations. Carambola juice is used to remove stains, because it contains oxalic acid.

References

Aranguren, C., C. Vergara, and D. Rosselli. "Toxicity of Star Fruit (*Averrhoa carambola*) in Renal Patients: A Systematic Review of the Literature." *Saudi Journal of Kidney Diseases and Transplantation*, vol. 28, no. 4, July/August 2017, p. 709–715.

"Ask Men's Health." *Men's Health*, vol. 21, no. 6, July/August 2006, p. 30.

"*Averrhoa carambola* L." USDA, NRCS. 2018. The PLANTS Database. National Plant Data Team, Greensboro, NC 27401-4901. https://plants.sc.egov.usda.gov/core/profile?symbol=AVCA. Accessed 07/12/18.

"Carambola." California Rare Fruit Growers. 1996. www.crfg.org/pubs/ff/carambola.html. Accessed 04/16/16.

"Carambola." University of Florida, Institute of Food and Agricultural Sciences Extension. http://gardeningsolutions.ifas.ufl.edu/plants/edibles/fruits/star-fruit.html. Accessed 08/24/17.

"Carambola: General Crop Information." University of Hawaii Extension. http://www.extento.hawaii.edu/kbase/crop/crops/i_caramb.htm. Accessed 07/13/18.

"Carambola. The Star Fruit." University of California, Agriculture and Natural Resources. www.ucanr.edu/sites/alternativefruits/files/64316.doc. Accessed 07/13/18.

Colwyn, Kim. "Star Fruit: Great Taste, Looks to Kill, and a Cinch to Prepare, This Fruit Really Is a Star." *Better Nutrition*, vol. 67, no. 12, December 2005, p. 20–21.

Crane, Jonathan H. "Carambola Growing in the Florida Home Landscape." HS12. University of Florida, Institute of Food and Agricultural Sciences Extension. November 2018. http://edis.ifas.ufl.edu/pdffiles/MG/MG26900.pdf. Accessed 07/11/18.

Dasgupta, P., P. Chakraborty, and N.N. Bala. "*Averrhoa carambola*: An Updated Review." *International Journal of Pharma Research and Review*, vol. 2, no. 7, July 2013, p. 54–63.

"Datasheet: Averrhoa carambola (carambola)." Centre for Agriculture and Biosciences International, January 3, 2018. www.cabi.org/isc/datasheet/8082. Accessed 07/10/18.

"Fresh Talent." *Good Health* (Australia Edition), April 2013, p. 90–92.

Gould, Walter P. "ARS Method Wins Ok to Ship Florida Carambola to Japan." *Agricultural Research*, vol. 43, no. 12, December 1995, p. 21.

Gunnars, Kris. "Star Fruit 101—Is It Good FOR You?" *Healthline*, March 24, 2016. www.healthline.com/nutrition/star-fruit-101. Accessed 07/10/18.

Lim, T.K. "Carambola." *Agnote*, 637, no. D27, March 1996. https://dpir.nt.gov.au/__data/assets/pdf_file/0008/233756/663.pdf. Accessed 07/13/18.

Mavin-Thomson, Leanne. "To Market Carambola." *The Newcastle Herald (includes the Central Coast Herald)*, October 31, 2007, Good Taste section, p.38.

McMahon, Gerry. "Fact Sheet: Carambola." FF2. Northern Territory Government (Australia). 2006. https://dpir.nt.gov.au/__data/assets/pdf_file/0006/227778/ff2_carambola.pdf. Accessed 07/16/18.

Morton, Julia F. *Fruits of Warm Climates*. Eugene, OR: Wipf and Stock Publishers, 2003.

Mullins, Lynne. "Carambolas and Sappodillas." *Sydney Morning Herald*, October 7, 2008, Good Living section, p. 11.

Nagy, Steven, editor, et al. *Fruits of Tropical and Subtropical Origin: Composition, Properties and Uses.* Lake Alfred, FL: Florida Science Source Inc., 1990.

Ortho All About Citrus and Subtropical Fruits. Des Moines, IA: Meredith Books, 2008.

Orwa, C., et al. "*Averrhoa carambola* L." 2009 Agroforestree Database version 4.0. www.worldagroforestry.org/sites/treedbs/treedatabases.asp. Accessed 07/14/18.

Piper, Jacqueline M. *Fruits of South-East Asia: Facts and Fiction.* Oxford, NY: Oxford University Press, 1989.

Popenoe, Wilson. *Manual of Tropical and Subtropical Fruits: Excluding the Banana, Coconut, Pineapple, Citrus Fruits, Olive and Fig.* NY: Macmillan Co., 1920.

Prag, Elliot, and Amy Spitalnick. "Star fruit." *Vegetarian Times*, no. 390, December 2011, p. 15.

"Starfruit (Carambola)." Plant Village, Pennsylvania State University. https://plantvillage.psu.edu/topics/starfruit-carambola/infos. Accessed 07/11/18.

"Tempting Tropicals: A Panoply of Antioxidants." *Prevention*, vol. 47, no. 4, April 1995, p. 104.

Todd, Kaley. "Environmental Nutrition: Let Star Fruit Shine." *Chicago Tribune*, July 6, 2015. www.chicagotribune.com/lifestyles/food/sns-201507061730—tms—foodstylts—v-f20150706-20150706-story.html. Accessed 07/10/18.

"Why You Should Avoid Eating Starfruit" National Kidney Foundation. www.kidney.org/atoz/content/why-you-should-avoid-eating-starfruit. Accessed 07/10/18.

Woods, Marcia. "Hey! Carambola." *Agricultural Research*, vol. 43, no. 3, March 1995, p. 22.

Cempedak (*Artocarpus integer* [Thunb.] Merr.)

In the same botanical family as the jackfruit, the cempedak somewhat resembles a scaled down version of a jackfruit. The size of the cempedak generally measures from 10 to 16 inches long and 4 to 6 inches in diameter, weighing less than 13 pounds. Its shape is compared to that of an oblong watermelon. The cempedak is also known as a small jackfruit, chempedak, and champadak.

The skin is hard and green when immature. Its skin is covered with blunt spines. The skin turns yellowish brown when ripe. When the fruit is ripe it emits an unpleasant odor, but not as strong and offensive as that of a durian. Most of the odor is supposedly removed once the cempedak's skin is removed. The fruit's flesh is yellow and its texture is often compared to custard. The flesh is arranged in segments. Each segment contains seeds measuring about 1 inch to 1¾ inches. The seeds are edible when cooked.

Unripe cempedaks have a taste similar to a sweet potato. Ripe cempedaks are similar to ripe jackfruits with a flavor like a blend of banana and pineapple. But it's sweeter and less acidic than the jackfruit.

CEMPEDAK LORE. Like the jackfruit, the cempedak is said to be a treatment for diabetes but there is currently an insufficient amount of studies to determine its effectiveness for this purpose.

TREE HISTORY. The tree originated in the Malay Peninsula and is widely grown in Malaysia, Indonesia, Thailand, Vietnam and India. The tree was

Cempedak *(EquatorialSky at English Wikipedia)*

also introduced into Australia, Jamaica, Kenya, Tanzania and Uganda and Hawaii.

TREE CHARACTERISTICS. In its native environment, the cempedak is a tropical forest tree. It's considered an understory tree, meaning it's a smaller tree that grows below the taller trees in the forest. The cempedak tree is an evergreen tree that can grow more than 60 feet tall. It has a dense rounded crown. And its bark is grayish brown to dark brown. Its leaves are dark green, oval, up to 9 inches long and four and three-quarters inches wide, and considered somewhat leathery.

The tree produces both male and female blossoms. The male blossoms are scented. The blossoms are creamy white and primarily wind pollinated. All parts of the tree contain a sticky milky latex.

Traditionally the trees have been propagated by seed. But trees are also propagated by grafting onto seedling rootstocks. And although trees propagated by grafting are truer to type, because of the sticky latex, grafting is more problematic. Trees grown from seed begin to fruit in three to six years. Grafted trees begin to fruit in two to four years.

In Malaysia a number of cultivars have been developed, notably those with orange flesh. Although primarily grown in Southeast Asia, cempedak is also grown in Australia and India. Unlike the jackfruit, cempedak is seldom grown in the United States.

Cempedak tree *(Tu7uh/Wikimedia Commons)*

▶ FAMILY: Moraceae. GENUS: *Artocarpus* J.R. Forst. & G. Forst. SPECIES: *Artocarpus integer* (Thunb.) Merr.

SELECTING A TREE. When selecting a tree from the nursery, check the tree's overall health. Make sure its leaves are not discolored, and free from insects or signs of insect damage. Check the entire tree to make sure it is free from injury, such as broken branches, cuts in the trunk, etc. Check to make sure the tree trunk is not sunburned. If possible, lift the tree out of its pot to inspect the roots. If the roots are serious girdled, you should avoid the tree because it is likely to develop a poor root structure.

HOW TO GROW A CEMPEDAK TREE. Cempedak trees grow best in well draining soils rich in organic matter. They prefer soils with a low pH. Mature trees produce best when full sun is available. To plant a tree, select a sunny location that can accommodate the size of a mature tree. Dig a hole at least twice the diameter of the pot the tree came in, and at least as deep as the container. Place the tree in the hole, filling the hole with the original soil. The tree should be planted so that the top of the root ball is about one to two inches higher than the soil level/grade. Water thoroughly.

Because of the thin weak stems of young cempedak trees, they should be staked their first year or two to prevent any potential wind damage. To properly stake a tree, place two stakes in the ground. Each stake should be placed on opposite sides of the tree, making sure they are placed outside of the root ball. Next, determine the proper height for the support ties. To do this, use your fingers and move it up the tree until the tree stands upright. Note the location, since the ties should be placed six inches above that point. If you have old pantyhose around, they make good ties since they provide the support needed, but also allow for any trunk growth.

The stakes and ties should be removed when the tap roots have grown enough to support the tree, or the tree is sturdy enough to withstand strong winds. Usually the stakes can be removed after a year or two. Make sure to remove the stakes when they are no longer needed, since leaving them in place can ultimately harm the tree with the ties and stakes causing tree injury, interfering with branch development, and stifling normal trunk growth.

Pruning is only needed to remove dead wood and if you want to control the height of the tree. It's recommended that mature trees be fertilized with N-P-K twice a year. The only reported major pest of the cempedak is the fruit fly.

HOW TO GROW FROM SEED. Cempedak seeds are recalcitrant and lose their viability when dried. So they should be planted shortly after removing a seed from the fruit. After removing seeds from the fruit, they should be cleaned and any fruit pulp washed away. Soak the seed in water for 24 hours. Then plant the seeds in moist soil. A minimum temperature of 80 degrees Fahrenheit is required for germination, which usually occurs within eight weeks.

▶ USDA HARDINESS ZONES: 10–11. CHILLING HOURS: 0. WATER REQUIREMENTS: Slightly higher than average.

HOW TO CONSUME. To eat ripe cempedaks fresh, lay the fruit on a tray lined with waxed paper or some other liner to prevent the fruit's sticky latex from dripping onto your counter. Use a knife and cut it from one end to the other. Pull the fruit open and remove each of the fruit segments from the skin. Next remove the seed from each of the segments. Enjoy the fresh pulp. The fresh pulp can also be pureed and mixed with ice cream or frozen and enjoyed as a dessert. The fresh pulp is commonly deep fried into fritters. Cempedak is also used in desserts like cakes and pudding.

To eat the seeds, they must first be cooked. Simply boil them for 30 minutes or until they are soft enough to be pierced by a fork. Remove from heat and drain. Remove and discard the seed shell and enjoy.

To roast the seeds, lay them out in a single layer on a baking pan. Roast for a minimum of 20 minutes at 400 degrees. Let cool, then remove the outer seed shell and discard it.

ADDITIONAL INFORMATION. Some studies have been done looking into the potential antimalarial properties of the tree bark. The tree leaves and ripe fruit are used as cattle feed. The timber is used in making furniture.

REFERENCES

"Artocarpus integer." Ecocrop Database. Food and Agriculture Organization of the United Nations. http://ecocrop.fao.org/ecocrop/srv/en/cropView?id=3435. Accessed 07/07/18.
"Artocarpus integer—(Thunb.) Merr." Plants for a Future. https://www.pfaf.org/USER/Plant.aspx?LatinName=Artocarpus+integer. Accessed 07/03/18.
"*Artocarpus integer* (Thunb.) Merr." USDA, NRCS. 2018. The PLANTS Database. National Plant Data Team, Greensboro, NC 27401-4901. https://plants.sc.egov.usda.gov/core/profile?symbol=ARIN17. Accessed 07/03/18.
"Cempedak (*Artocarpus integer*)." Infopedia. National Library Board, Singapore. http://eresources.nlb.gov.sg/infopedia/articles/SIP_201_2005-02-01.html. Accessed 07/05/18.
"Cempedak (*Artocarpus integer*)." National Gardening Association Plants Database. https://garden.org/plants/view/209287/Cempedak-Artocarpus-integer/. Accessed 07/07/18.
Cempedak—A Close Cousin of the Jackfruit. Kanjirappally, India: Biotec Homegrown. n.d.
Chandlee, David K. "Chempedak—Artocarpus integer." Subtropical Fruit Club of Queensland. February 2006. http://stfc.org.au/chempedak-artocarpus-integer. Accessed 07/07/18.
"Chempedak, Also Called Small Jackfruit *Artocarpus integer*" Rare Fruit Club, Western Australia. April 8, 2014. www.rarefruitclub.org.au/Level2/Chempedak.htm. Accessed 07/08/18.
Hafid, Achmad Fuad, et al. "The Active Marker Compound Identification of *Artocarpus Champeden* Spreng. Stembark Extract, Morachalchone A as Antimalarial." *International Journal of Pharmacy and Pharmaceutical Sciences*, vol. 4, suppl. 5, 2012, p. 246–249.
Jensen, M. *Trees Commonly Cultivated in Southeast Asia: An Illustrated Field Guide.* Food and Agriculture Organization of the United Nations, Regional Office for Asia and the Pacific. April 2001. www.fao.org/docrep/005/ac775e/ac775e03.htm. Accessed 07/07/18.
Morton, Julia F. *Fruits of Warm Climates.* Eugene, OR: Wipf and Stock Publishers, 2003.
Orwa, C., et al. "Artocarpus integer." 2009 Agroforestree Database: A Tree Reference and Selection Guide. www.worldagroforestry.org/sites/treedbs/treedatabases.asp. Accessed 07/03/18.
Pui, L.P., et al. "Physicochemical and Sensory Properties of Selected 'Cempedak' (*Artocarpus integer* L.) Fruit Varieties." *International Food Research Journal*, vol. 25, no. 2, April 2018, p. 861–869.
Wang, Maria, et al. "Origin and Diversity of an Underutilized Fruit Tree Crop, Cempedak (Artocarpus integer, Moraceae)" *American Journal of Botany*, vol. 105, no. 5, May 2018, p. 813–957.

Che (*Cudrania tricuspidata* Bur. ex Lavallee)

Also known as a Chinese melonberry, a Mandarin melonberry, Cudrang, and a Chinese mulberry, the che fruit measures about one to two inches in diameter. It has a lumpy surface like a raspberry that turns red or purplish red when ripe. A fully ripe che features a soft juicy red pulp with three to six small brown edible seeds in the middle. Its taste is frequently described as very mild, sweet, and similar to a watermelon with a pinch of cotton candy and fig mixed in.

CHE LORE. In Asian herbal medicine, che is said to prevent diabetes, cancer and heart disease. It's also supposed to be good for weight loss. However there is a lack of research to support its use for these conditions.

TREE HISTORY. Originating in eastern Asia from China to the Himalayas, the tree spread to Japan where it became naturalized. Because silkworms feed on the tree's leaves, the tree is also known as a Silkworm Thorn. Some believe it was the silk trade that helped spread the tree from Asia, to Europe, and other parts of the world. The tree was introduced into France in 1862, into England around 1872, and into the U.S. in the early 1900s.

TREE CHARACTERISTICS. Growing 15 to 25 feet tall, the che is a deciduous small tree or bush. It will become a spreading bush about as wide as it is tall, if left to grow naturally. Young tree branches have thorns. It bears varying alternate pale yellowish green leaves that can have three lobes or no lobes at all. The trees are generally female or male trees, meaning you'll need both

Che *(Assianir/Wikimedia Commons)*

Unripe che on plant *(Krzysztof Ziarnek, Kenraiz/Wikimedia Commons)*

a male and female trees for good fruiting, although some trees have been known to produce both male and female blossoms. The blossoms are tiny and white or green, with male blossoms turning yellow as they release their pollen. The blossoms are pollinated by the wind. Mature trees are fairly drought tolerant and wind resistant. Mature trees can produce several hundred pounds of fruit annually.

The trees are propagated by seed, by softwood cuttings, and grafting. Che is frequently grafted onto Osage orange (*Maclura pomifera*) root stock. Trees grown from seed can take ten years before they begin to fruit, whereas trees propagated from cutting and grafting will begin to fruit significantly sooner.

▶ FAMILY: Moraceae. GENUS: *Cuadrania* Trecul. SPECIES: *Cudrania tricuspidata* (Carrière) Bureau ex Lavallée.

SELECTING A TREE. When selecting a tree from the nursery, check the tree's overall health. Look for active growth, such as new or young leaves.

Make sure its leaves are not yellow or discolored, and the tree is free from insects or signs of insect damage. Check the entire tree to make sure it is free from injury, such as broken branches, cuts in the trunk, etc. And if possible, gently lift the tree out of its pot to check the roots. If the roots are overgrown in the pot and are seriously girdled, the tree should be avoided since it will likely develop a poor root structure.

How to Grow a Che Tree. Che trees grow best in a sunny location and in a wide range of soils. But they grow best in well draining loam soils. They prefer a soil pH between 6.1 to 6.5 but will tolerate up to 7.8. And remember you will need to plant both a male and female tree for good fruiting.

Once you've selected the best location for your trees, dig a hole at least twice the diameter of the pot the tree came in and about as deep as the container. Place the tree in the hole, loosening or cutting away any girdled roots. The tree should be planted so that the top of the root ball is about an inch or two higher than the soil level. Refill the hole with the original soil. Water thoroughly. The tree should be watered regularly until it becomes established. Trees should also be watered regularly during the summer for best fruiting. For best fruiting, the tree requires regular pruning since it fruits on new wood. Prune back the previous season's growth by half.

How to Grow from Seed. To grow from seed, remove the seed from a ripe che fruit and wash away any pulp. Plant the seed in a moist planting medium in a pot. As a general rule, plant seeds as deep in the soil as the size of the seed. Place the pot in a warm location and keep the soil moist until the seed germinates. It's recommended the seedling be grown in a greenhouse during the first winter, and then planted outside in late spring after all danger of frost has passed.

USDA Hardiness Zones: 5–10. Chilling Hours: —. Water Requirements: Moderate.

How to Consume. Immature che fruit are said to be tasteless. Ripe fruit are very soft to the touch and can be eaten fresh. Fresh fruit can be stored in a sealed container for several days. The fruit is also often turned into a juice by mixing it in a blender and then straining out the seeds. The juice can be drunk as is, or mixed with other fruit and made into a smoothie. Che are also used to flavor sorbets, and used to make cobblers and tarts.

Additional Information. In Japan, the che tree is used in the art of bonsai.

REFERENCES

Burrell, Beth. "Chinese Che Tree." *Virginia Gardener,* vol. 9, no. 9, Dec. 2, 2011. http://state bystategardening.com/state.php/site/articles/chinese_che_tree/. Accessed 06/07/18.

"Che." California Rare Fruit Growers. http://www.crfg.org/pubs/ff/che.html. Accessed 10/22/15.

"Che." Growable: Grow Florida Edibles. http://www.growables.org/information/TropicalFruit/che.htm. Accessed 10/22/15.

"Che." Specialty Produce. www.specialtyproduce.com/produce/Che_Fruit_11146.php. Accessed 06/07/18.

"Che Fruit." Frutalestropicales. http://frutalestropicales.com/en/home/178-che-fruit-maclura-tricuspidata.html. Accessed 06/07/18.

"Chinese Mulberry." Grow Plants. https://www.growplants.org/growing/chinese-mulberry. Accessed 12/21/18.

"*Cudrania tricuspidata*—(Carrière.) Bur. ex Lav." Plants for a Future. https://pfaf.org/User/Plant.aspx?LatinName=Cudrania+tricuspidata. Accessed 12/19/18.

"*Cudrania tricuspidata* (Carrière) Bureau ex Lavallée." USDA, NRCS. 2018. The PLANTS Database. National Plant Data Team, Greensboro, NC 27401-4901. https://plants.sc.egov.usda.gov/core/profile?symbol=CUTR2. Accessed 12/18/18.

"Melonberry." Plant Database. https://plantdatabase.earth/melonberry. Accessed 12/21/18.

Reich, Lee. "Che: Chewy Dollops of Maroon Sweetness." *Arnoldia: A Continuation of the Bulletin of Popular Information of the Arnold Arboretum, Harvard University,* vol. 64, no. 1, 2006, p. 31–36. http://arnoldia.arboretum.harvard.edu/pdf/articles/2006-64-1-che-chewy-dollops-of-maroon-sweetness.pdf. Accessed 12/21/18.

Reich, Lee. *Uncommon Fruits for Every Garden.* Portland, OR: Timber Press. 2004.

"Seedless Che Fruit Care Guide." Edible Landscaping. https://ediblelandscaping.com/careguide/SeedlessCheFruit/. Accessed 12/21/18.

Stallsmith, Audrey. "Uncommon Fruit Trees for Adventurous Gardeners." Dave's Garden. January 35, 2013. https://davesgarden.com/guides/articles/view/4065. Accessed 12/21/18.

Trevino, Laramie. "PLANT OF THE WEEK: Che / Che, or Chinese melonberry, grows best with lots of heat / Bay Area fruit tends to be smaller, not as sweet, but still appealing." SFGate. www.sfgate.com/homeandgarden/article/PLANT-OF-THE-WEEK-Che-Che-or-Chinese-2585289.php. Accessed 10/22/15.

Weydemeyer, Idell. "Grafting Dormant Deciduous Fruit Scions." Golden Gate Chapter, California Rare Fruit Growers. http://ccmg.ucanr.edu/files/172574.pdf. Accessed 12/21/18.

Chinese Quince (*Pseudocydonia sinensis* [Thouin] C.K. Schneid)

Not to be confused with the common quince, the Chinese quince, also known as a false quince, is an aromatic large oval fruit averaging five to seven inches long and about four to six inches in diameter. It can weigh up to two pounds. It turns bright yellow when mature, and is loosely similar to a pear in appearance. Its pulp is somewhat hard with a slight sandy texture. It has several large seeds, larger than those in a common quince. It is tart, acidic and astringent and not eaten fresh. Its taste, when jellied, is compared to a cross between an apple and a pear.

CHINESE QUINCE LORE. Quince has been used in folk medicine to treat a variety of conditions ranging from coughs to diarrhea, digestive problems,

Chinese quince *(Tahir mq/Wikimedia Commons)*

and inflammation. However, there are currently insufficient medical studies to prove the effectiveness of quince for treatment of these medical conditions.

TREE HISTORY. The quince may have been the "forbidden fruit" in the Bible. The old Biblical name for quince translates into "golden apple" and predates the growing of apples in Mesopotamia. Eve may have taken a bite out of a quince.

Originally classified as *Cydonia sinensis*, the Chinese quince originated in China. It was first recorded during the Tang Dynasty (618–907). It was introduced into the United States around 1800. It was later introduced into England and Holland, and then France.

TREE CHARACTERISTICS. The Chinese quince is a deciduous small tree or large shrub; it generally grows 10 to 20 feet tall, but can grow up to 30 or 40 feet tall in optimal conditions. Its width or spread is only slightly smaller (about a third less) than its height. It has oval to elliptic shaped glossy dark green leaves measuring two and a half to four and a half inches long and one and a half to two and a half inches wide with finely serrated edges. The leaves turn shades of yellow and red in the fall.

In the spring the tree bears pale pink blossoms. The single blossoms are one to one and a half inches wide and fragrant. The blossoms are pollinated by bees and other insects.

The tree's exfoliating bark is considered to be one of the most beautiful tree bark patterns. On mature trees pieces of bark are shed or peel away exposing a mosaic of grey, green, tan and orange patches on the tree's fluted trunk.

Chinese quince bonsai tree *(Sage Ross/Wikimedia Commons)*

Chinese quince trees are propagated by layers, cuttings, and grafting. They can also be propagated by seed, although trees grown from seed may not grow true to type. Trees grown from seed usually begin bear fruit in their fifth year.

Popular quince cultivars include the 'Pineapple' which has a pineapple flavor and white pulp, the 'Orange,' which has a rich flavor and orangey yellow pulp, and the 'Smyrna,' which produces large fruit.

▶ FAMILY: Rosaceae. GENUS: *Pseudocydonia* C.K. Schneid. SPECIES: *Pseudocydonia sinensis* (Thouin) C.K. Schneid.

SELECTING A TREE. If you buy a three or four year old tree, the tree's basic shape has already been formed for you. When selecting a tree from the nursery, check the tree's overall health. Look for active growth, such as new or young leaves. Make sure its leaves are not yellow or discolored, and free from insects or signs of insect damage. Check the entire tree to make sure it is free from injury, such as broken branches, cuts in the trunk, etc. And if possible, gently lift the tree out of the pot to check the roots. Overgrown trees that have seriously girdled roots will likely develop a poor root structure so they should be avoided.

HOW TO GROW A CHINESE QUINCE TREE. The trees grow best in full sun, preferring slightly acidic, well draining loam soils with even moisture. They will grow in sandy loam to some clay soils, and even tolerate poor soil. It will grow in soil with a pH range from 5.5 to 7.5. You should plant two trees for good pollination.

The best time to plant a quince tree is in the spring after all danger of frost has passed. To plant a tree, select a location that receives at least six hours of sun daily that can accommodate the size of a mature tree. Dig a hole at least twice the diameter of the pot the tree came in, about as deep as the container. Place the tree in the hole, loosening or cutting away any girdled roots. The tree should be planted so that the top of the root ball is about one to two inches higher than the soil level. Refill the hole with the original soil. Water the newly planted tree thoroughly. Do not irrigate the tree during bloom since it increases the risk of fire blight. If you want to encourage root development, young trees can be supplied with some extra phosphorus when they're planted, or during the first year it is planted.

The trees require little pruning. Pruning is only needed to remove dead branches, and remove overcrowded branches. You want to make sure light reaches the interior of the canopy for good fruit development. Pruning is also required if you want to keep the tree a certain size.

The Chinese quince is susceptible to fire blight, a bacterial disease caused by *Erwinia amylovora*. Fire blight is more likely to occur when the temperature is between 75 and 85 degrees Fahrenheit followed by intermittent rain or hail.

Signs of fire blight include a watery ooze from twigs, branches, and trunk cankers. Wilted black blossoms, blackened fruit, and localized branch wilts are also signs. If the disease has spread into the tree wood, the wood underneath the bark will have pink to orange red streaks. And if infected bark is removed or cut away, you'll see reddish flecking near the canker.

If your tree is infected, remove and destroy any infected areas in the summer or winter when the disease is no longer actively spreading throughout the tree. After removing diseased areas, be sure to disinfect your pruning tools so you won't spread the disease to other plants. (To disinfect your tools, dip them in a solution of 10 percent bleach and 90 percent water.)

For more information on fire blight and how to remove diseased areas from major limbs and tree trunks, you can consult the Fire Blight Management Guidelines from the University of California accessible online at http://ipm.ucanr.edu/PMG/PESTNOTES/pn7414.html.

How to Grow from Seed. Remove the seeds from ripe quinces. Remove any pulp from the seeds and wash the seeds, drying them with a paper towel. The seeds need to be stratified. Place the seeds in a zipper sealed sandwich bag filled about three-quarters full with sand and place the bag in your refrigerator. In the spring, remove the seeds from the refrigerator. Plant them about half an inch deep in moist potting soil. Keep the soil moist, but not wet. The seeds should germinate in about six weeks.

▶ USDA Hardiness Zones: 6–9. Chilling Hours: 300. Water Requirements: Moderate.

How to Consume. Because they are astringent, Chinese quinces are not eaten fresh. Quinces are frequently made into jellies, jams, marmalades, and other preserves. They are high in pectin. Quinces can be baked into pies or tarts with apples, or used in other baked goods. They can be poached and served with ice cream. Chinese quinces are frequently made into a paste. They can be stewed and are frequently used to make liqueur. The National Center for Home Food Preservation recommends quinces NOT be dehydrated.

Additional Information. WARNING: If you are taking any oral medications, you should wait at least an hour after taking your medication before consuming a quince. The quince has a fiber that decreases your body's absorption of your medications.

Because of their pectin content, one of the early uses of quinces was for marmalade. The Portuguese word for quince is marmelo. It later became marmalade, a name originally given to a conserve made with quinces. Quince seedlings are often used as rootstock for pear trees.

REFERENCES

Andress, Elizabeth L., and Judy A. Harrison. *So Easy to Preserve*. 6th ed. Bulletin 989. Athens, GA: College of Family and Consumer Sciences, College of Agricultural and Environmental Sciences, Cooperative Extension, University of Georgia. 2014.

Balge, Russell. "Barking Up the Right Tree." *American Nurseryman*, vol. 183, no. 7, April 1, 1996, p. 44–49.

"Can I Grow Quince Trees From Seed: Learn About Quince Seed Germination." Gardening Know How. www.gardeningknowhow.com/edible/fruits/quince/quince-tree-seed-propagation.htm. Accessed 10/06/18.

"Cydonia sinensis Thouin." Atlas of Living Australia. https://bie.ala.org.au/species/NZOR-4-30270. Accessed 10/01/18.

"Fire Blight." University of California Statewide Integrated Pest Management Program. http://ipm.ucanr.edu/PMG/PESTNOTES/pn7414.html. Accessed 10/06/18.

Higgins, Adrian. "The Rare Beauty of the Chinese Quince." *The Washington Post*, December 9, 2010. www.washingtonpost.com/wpdyn/content/article/2010/12/07/AR2010120706002_pf.html. Accessed 09/30/18.

Hill, Lewis, and Leonard Perry. *The Fruit Gardener's Bible: A Complete Guide to Growing Fruits and Nuts in the Home Garden*. North Adams, MA: Storey Publishing, 2011.

"How to Grow Quince Seed." Garden Guides. www.gardenguides.com/89914-grow-quince-seed.html. Accessed 10/06/18.

Ingels, Chuck, et al., editors. *The Home Orchard: Growing Your Own Deciduous Fruit and Nut Trees*. Richmond, CA: University of California, Agriculture and Natural Resources, 2007.

Marshall, Kate. "How to Grow Quinces." *Taranaki Daily News*, April 23, 2016, p. 26.

National Center for Home Food Preservation. https://nchfp.uga.edu/. Accessed 04/03/17.

Newman, Jacqueline M. "Quince." *Flavor and Fortune*, vol. 6, no. 3, Fall 1999, p. 13–14.

Pittenger, Dennis. *California Master Gardener Handbook*. 2nd ed. University of California, Agricultural and Natural Resources. 2014.

Postman, Joseph. "Cydonia oblonga: The Unappreciated Quince." *Arnoldia*, vol. 67, no. 1, 2009, p. 2–9.

"*Pseudocydonia sinensis* (Chinese Quince)." The Backyard Gardener. www.backyardgardener.com/plantname/pseudocydonia-sinensis-chinese-quince/. Accessed 09/30/18.

"*Pseudocydonia sinensis*." Missouri Botanical Garden. www.missouribotanicalgarden.org/PlantFinder/PlantFinderDetails.aspx?kempercode=d455. Accessed 09/30/18.

"*Pseudocydonia sinensis*." Plant Database. University of Connecticut, College of Agriculture, Health and Natural Resources. http://hort.uconn.edu/detail.php?pid=367. Accessed 09/30/18.

"*Pseudocydonia sinensis* (Thouin) C.K. Schneid." USDA, NRCS. 2018. The PLANTS Database. National Plant Data Team, Greensboro, NC 27401-4901. https://plants.sc.egov.usda.gov/core/profile?symbol=PSSI4. Accessed 09/30/18.

"*Pseudocydonia sinensis* (Thouin)." Flora of North America, vol. 9. http://efloras.org/florataxon.aspx?flora_id=1&taxon_id=220011075. Accessed 10/01/18.

"Quince." University of California, Davis, College of Agricultural and Environmental Sciences, Fruit and Nut Education Center. http://fruitandnuteducation.ucdavis.edu/fruitnutproduction/Quince/. Accessed 10/05/18.

"Quince." WebMd. www.webmd.com/vitamins/ai/ingredientmono-384/quince. Accessed 01/01/18

"Quince Chinese." Incredible Edibles. Tharfield Nursery, Ltd. www.edible.co.nz/fruits.php?fruitid=81_Quince%20Chinese. Accessed 06/15/18.

"The Story of the Quince." Prospect Books (London). n.d. https://prospectbooks.co.uk/wp-content/uploads/2014/09/Quinces_extract.pdf. Accessed 10/06/18.

USDA, Agricultural Research Service, National Plant Germplasm System. 2018. Germplasm Resources Information Network (GRIN-Taxonomy). National Germplasm Resources Laboratory, Beltsville, Maryland. https://npgsweb.ars-grin.gov/gringlobal/taxonomyde-tail.aspx?318589. Accessed 10/01/18.

Citron (*Citrus medica* L.)

The citron is a citrus species with many varieties. The citron is usually oblong or oval shaped, but there are many varieties taking different shapes and forms. It ranges from four to 11 inches long. Its rind is rough and usually somewhat lumpy. It generally turns from green to yellow when mature. Most citrons possess a thick white pith. They generally have a pale yellow pulp, although pulpless varieties exist. They are usually acidic to mildly acidic, although there are sweet varieties.

CITRON LORE. According to Richard Folkard's *Plant Lore, Legends and Lyrics,* "During the feast of the Tabernacles, the Jews in their synagogues carry a Citron in their left hand; and a conserve made of a particular variety of the fruit is in great demand by the Jews, who use it during the same feast. According to Athenæus, certain notorious criminals, who had been condemned to be destroyed by serpents, were miraculously preserved, and kept in health and safety by eating Citrons. Theophrastus says that Citrons were considered an antidote to poisons, for which purpose Virgil recommended them in his Georgics."

Various parts of the fruit and trees have been used through the ages in folk medicine. The citron peel has been used as an expectorant, and also to treat dysentery. The juice has been used as both a sedative and as a treatment to purge poison from a person's system. The tree's leaves are also used as a treatment for poison, as well as stomach and intestinal problems. And the fruit has been used to treat everything from rheumatism to motion sickness.

TREE HISTORY. The citron is believed to have originated in Southeast Asia, or in the westernmost area of Asia in the central Himalayan foothills. The citron dates back to at least 4000 BCE. Botanists found evidence of the citron in Nippur, an ancient city in Mesopotamia in an area now part of South Eastern Iraq. The citron was cultivated in the Persian Gulf in ancient times. After his attack on Persia, some historians credit Alexander the Great with spreading the citron to the Mediterranean basin and Greece, Italy and other European countries. According to palynological (pollen and spores) evidence, the citron was the first citrus species to migrate west from Asia.

It's believed the citron was introduced into the United States by the Spaniards in the late 1800s. Several early attempts to commercially cultivate

Citron *(sarangib/pixabay.com)*

citrons in California and Florida failed because of frost damage. But that has changed and today citrons are produced primarily in California and Florida.

TREE CHARACTERISTICS. Citron plants are slow growing small trees or shrubs usually growing about 8 to 15 feet in height. The tree's evergreen leaves are ovate elliptic and measure from about two and a half to seven inches long. Young leaves also have a slight lemon scent. The tree bears stiff straggly thorny branches. Citron trees bear blossoms in small clusters. The blossoms are large, white or purple tinted. Measuring about one and a half inches wide, they are also fragrant. The blossoms are pollinated by bees and other insects.

Citron trees can be propagated by seed, by cuttings, and budding. Most trees sold commercially are propagated by cuttings or by budding, because those grown from seed may not bear true to type. Trees begin fruiting when they are two to three years of age.

Mature trees can produce from 80 to more than 200 pounds of fruit annually. There are countless citron varieties and cultivars. They run the gamut from sweet to acidic pulp, to pulpless cultivars. Some have thin peels and others thick peels, some with seeds and some without.

► FAMILY: Rutaceae. GENUS: *Citrus* L. SPECIES: *Citrus medica* L.

SELECTING A TREE. Due to the numerous varieties and cultivars, be sure to select a citron variety or cultivar for your specific needs. When selecting a tree from the nursery, check the tree's overall health. Look for active growth, such as new or young leaves. Make sure its leaves are not yellow or discolored, and free from insects or signs of insect damage. Check the entire tree to make sure it is free from injury, such as broken branches, cuts in the trunk, etc. And if possible, gently lift the tree out of the pot to check the roots. Overgrown trees that have seriously girdled roots will likely develop a poor root structure so they should be avoided.

HOW TO GROW A CITRON TREE. Citron trees tolerate a wide range of soils providing there's adequate drainage. But they prefer well draining, moisture retaining loamy soils. They prefer full sun. The best time to plant a citron tree is in the spring when all danger of frost has passed. Select a sunny site shielded from strong winds that can easily accommodate the size of a mature tree. Dig a hole as deep as the container the tree came in, and twice the width. When you place the tree in the hole, the root ball should sit about one to two inches above the soil. Refill the hole with the original soil. Water thoroughly. The tree should be watered regularly to help it become established. Once it is established, it needs water to produce decent sized juicy fruit. But don't over water the tree because it makes it more susceptible to root rot.

Young citron trees are sensitive to frost. If temperatures are expected to dip to 29 degrees Fahrenheit or lower, you need to protect the tree from frost.

Candied citrus peel (*pixabay.com*)

You should wrap the tree trunk and branches in insulation material like cardboard, fiberglass or frost protection covers. Make sure the soil is moist and not dry, since damp soil retains and radiates more heat than dry soil. Bare soil also radiates more heat than soil covered with mulch or other ground covers.

Very little pruning is required. Any suckers growing from below the bud/graft union should be removed. You should also prune the tree if you want to keep it a certain height to make it easier to harvest. Citron trees, like other citrus trees, can also be espaliered. Mature trees will benefit from a regular application of nitrogen prior to blossoming in January or February, and then a second application in May or June.

How to Grow from Seed. To grow a tree from seed, remove the seeds from the citron fruit. Wash any pulp away from the seeds. Soak the seeds in clean water overnight. Place the seeds about half-inch deep in a sterile moist planting/potting medium. Cover the pot with plastic to create a humid atmosphere. Keep the soil moist until seedlings sprout. The seeds germinate best in temperatures above 60 degrees Fahrenheit. The seeds usually germinate in three to four weeks. Once the seeds sprout, transplant them into a larger pot.

▶ USDA Hardiness Zones: 8–10. Chilling Hours: 0. Water Requirements: Average.

How to Consume. Those citron varieties and cultivars that have a juicy pulp are most frequently juiced. The juice is often used as a substitute for lemon juice. The juice is also infused into liqueurs. Vodka and gin are frequently used. To infuse it into alcohol, wash and slice the citron. Place the fruit into an airtight jar, like a canning jar. Pour the vodka or other liqueur over the fruit. Seal the jar and shake it a few times. On a daily basis shake the jar. Let the fruit infuse, i.e., let the jar sit for five days to two weeks, depending upon how strong you want the flavor to be. When it's ready, strain the fruit from the alcohol and enjoy.

As a citrus fruit, citron is often used in marmalades. The citron peel is also commonly candied. The candied peel can be eaten as is, or can be used in various baked goods, such as cakes, breads and cookies. There are several methods of candying the peel. Following is one method (be sure to remove any thick pith from the peel): Wash it thoroughly and pat it dry. Slice the peel into quarter-inch strips. In a pot, boil equal parts of sugar and water, for example two cups water and two cups sugar, stirring until the sugar is dissolved. Add in the peel strips and simmer for 20 to 45 minutes, or until the strips turn translucent. Remove the strips and place them on a wire rack to cool. Once they've cooled, lightly sprinkle the strips with granulated sugar so they are evenly coated.

Another method to candy the fruit is to boil the strips in boiling water for about 10 minutes until the peel is tender, then remove and place the strips on a wire rack for about 15 minutes. Then add the strips a boiling solution of equal parts water and sugar for 15 to 30 minutes, remove, cool, and coat with sugar.

ADDITIONAL INFORMATION. Because of the citron's fragrance, it has also historically been used as a perfume, and moth repellant.

REFERENCES

Brigand, Jean-Paul, and Peter Nahon. "Gastronomy and the Citron Tree (*Citrus medica* L.)." *International Journal of Gastronomy and Food Science*, vol. 3, April 2016, p. 12–16.

Chhikara, Navnidhi, et al. "*Citrus medica*: Nutritional, Phytochemical Composition and Health Benefits—A Review." *Food & Function*, vol. 9, no. 4, April 25, 2018, p. 1978–1992.

"Citron." Grow Plants. www.growplants.org/growing/citron. Accessed 12/04/18.

"Citron." KidzSearch. https://wiki.kidzsearch.com/wiki/Citron. Accessed 12/04/18.

"Citron." The Eden Project, Cornwall, UK. www.edenproject.com/learn/for-everyone/plant-profiles/citron. Accessed 12/04/18.

"Citrus—Citron." LEAF Network, AZ. https://leafnetworkaz.org/resources/PLANT%20PROFILES/Citrus_Citron_profile.pdf. Accessed 12/04/18.

"*Citrus medica*." Plantlives. www.plantlives.com/docs/C/Citrus_medica.pdf. Accessed 12/04/18.

"*Citrus medica*." *Useful Trees and Shrubs of Ethiopia*. www.worldagroforestry.org/usefultrees/pdflib/Citrus_medica_ETH.pdf. Accessed 12/04/18.

"*Citrus medica* L." USDA, NRCS. 2018. The PLANTS Database. National Plant Data Team, Greensboro, NC 27401-4901 USA. https://plants.sc.egov.usda.gov/core/profile?symbol=CIME3. Accessed 12/04/18.

"*Citrus medica* L. (Rutaceae)." Department of Plant Sciences, University of Oxford. https://herbaria.plants.ox.ac.uk/bol/plants400/profiles/CD/Citrusmedica. Accessed 12/04/18.

Fern, Ken. "*Citrus medica*." Tropical Plants Database. http://tropical.theferns.info/view tropical.php?id=Citrus+medica. Accessed 12/04/18.

Folkard, Richard. *Myths, Traditions, Superstitions and Folk Lore of the Plant Kingdom*. London: Sampson, Low, Marston, Searle and Rivington. 1884.

Klein, Joshua D. "Citron Cultivation, Production and Uses in the Mediterranean Region." www.researchgate.net/publication/300027232_Citron_Cultivation_Production_and_Uses _in_the_Mediterranean_Region. Accessed 12/04/18.

Langgut, Dafna. "The History of Citrus Medica (citron) in the Near East: Botanical Remains and Ancient Art and Texts." In: *AGRUMED: Archaeology and history of citrus fruit in the Mediterranean: Acclimatization, diversification, uses* [online]. Naples: Publications du Centre Jean Bérard, 2017. http://books.openedition.org/pcjb/2184. Accessed 12/05/18.

Meena, Ajay Kumar, et al. "A Review on Citron Pharmocognosy, Phytochemistry and Medicinal Uses." *International Research Journal of Pharmacy*, vol. 2, no. 1, January 2011, p. 14–19.

Morton, Julia F. *Fruits of Warm Climates*. Eugene, OR: Wipf and Stock Publishers, 2003.

Nicolosi, Elisabetta, et al. "The Search for the Authentic Citron (*Citrus medica* L): Historic and Genetic Analysis." *Hortscience*, vol. 40, no. 7, December 2005, p. 1963–1968. http://hortsci.ashspublications.org/content/40/7/1963.full.pdf+html. Accessed 12/04/18.

Ortho All About Citrus and Subtropical Fruits. Des Moines, IA: Meredith Books, 2008.

Rademaker, Marius. "Citron." Dermnet. Hamilton, New Zealand, 1999. www.dermnetnz.org/topics/citron. Accessed 12/04/18.

Sauls, Julian W. "Home Fruit Production—Miscellaneous Citrus." Texas Cooperative Extension. December 1998. https://aggie-horticulture.tamu.edu/citrus/miscellaneous.htm. Accessed 12/04/18.

Durian (*Durio zebethinus* Murray)

Also known as the "king of fruits" in Southeast Asia, durians are oblong to sometimes almost round shaped fruits measuring up to six inches in diameter and a foot long, about the size of a football. Although it can weigh up to 18 pounds, most weigh under eight pounds. The outer skin of the durian is olive green to yellow depending upon the cultivar. The outer skin is hard, and covered with pointed spikes that feel like thorns. Thus the fruit is best handled with gloves. Within the skin the pulp is segmented into five cavities. The custard-like pulp is generally cream colored or yellowish to orange or red depending upon the cultivar. The fruit has a handful of large seeds about the size of a chestnut.

People often describe durians as an acquired taste. Durian aficionados say it tastes like a blend of banana, caramel and butterscotch, or almond flavored custard. Others describe it as tasting like powdered sugar and whipped cream with a bit of chives mixed in. It's also often described as tasting like vanilla custard with a hint of garlic. Some people who taste the fruit for the first time, and others who find the taste offensive, describe it as tasting like a severely rotten onion.

Cooked durian seeds are edible. But the raw durian seeds contain cyclopropene fatty acids which can be toxic. Heat destroys the acids, making the seeds edible.

Durians are infamous for the odor they emit when ripe. Although some cultivars have less of an odor, durian fans prefer those with strong odors for their taste. It's said the odor can be smelled up to half a mile away. The odor has been compared to everything from rotten eggs to onions, turpentine and old gym socks. Their smell can be so offensive that durians have been banned from public transportation in Singapore, and other public places in various Southeast Asian countries.

DURIAN LORE. There's an abundance of lore about the durian. There's also no shortage of verbal legends about the durian. Although some do qualify as legends, a few are questionable as to their origins. One very popular legend has various versions told from one South East Asia country to another. But the basic story is the same. Basically, the legend is as follows:

There once lived a powerful old king who took a beautiful young woman as his wife. But she was not attracted to the king, so she kept running off to her father, who dutifully returned her to her husband. In his search to find a way to make his wife love him, the king consults a hermit. In some versions, the hermit is said to be a demigod, or have magical powers. The hermit tells the king he can help him, but he must gather three things—the egg of a black tabon bird, 12 ladles of milk from a white water buffalo, and the flower from

Durian *(Kalai/Wikimedia Commons)*

a magical tree (in some versions it's the Tree of Vision, in others it's the Tree of Make Believe). The king gathered the items and brought them to the hermit.

The hermit mixed the items together, gave the mixture to the king and told him to plant it in the palace garden. From it, the hermit said, would grow a tree with fragrant, delicious fruit with magical powers. As payment for his service, the hermit only asked that he be honored at a feast to celebrate his wife's new love for him. The king planted the mixture and from it grew a durian tree. When his wife ate a durian fruit, she suddenly saw the king as a young, handsome man and she fell madly in love with him. To celebrate, the king held a grand feast but forgot to invite the hermit. Angry at being forgotten, the hermit changed the pleasant smell of the durian to an odorous one and covered it with thorns.

Another legend about the origin of the durian tree concerns a great sultan and his beautiful wife. They had a daughter they named Duri, who had a kind heart but was born ugly. So ugly, she had no friends and no suitors as she grew up. Lonely and depressed, she got very sick. One night she heard singing and got out of bed and followed the singing. The next morning she was found dead. She was buried near her village. Years passed. Then a man

Cut durian *(pixabay.com)*

from a nearby village noticed a strange tree growing where Duri had been buried. The tree bore fruit with a prickly exterior and horrible smell. But curious about the fruit, he decided to taste it and discovered it was delicious. The people were surprised that something so hideous was full of such goodness, just like Duri. So they named the fruit Durian after her. In another version of this legend, Duri died from unrequited love. A durian tree grew where she was buried. The thorny surface of the durian is said to represent her pain of unrequited love, and the delicious flesh is said to represent her goodness.

It's said that the durian is an aphrodisiac. There's no scientific evidence to substantiate this. It's believed that this myth arose from the legend of how eating a durian led the wife of an old king to see him as young and handsome and fall in love with him.

There's a popular myth that eating a durian and drinking alcohol can kill you. Unfortunately, there may be some truth to this myth. An article in *New Scientist* reported that researchers from the University of Tsukuba, Japan, found that durian extract inhibited the activity of aldehyde dehydrogenase (enzymes that clear toxic breakdown of products and are crucial for alcohol metabolism) by up to 70 percent in a laboratory setting. So you may want to avoid drinking alcohol when eating durians.

TREE HISTORY. The tree is thought to have originated in the Malaysian rain forest. It is native to the islands of Borneo (in the countries of Malaysia, Brunei, and Indonesia) and Sumatra (in western Indonesia). Durian is grown across Southeast Asia and is believed to have been consumed there since prehistoric times. The earliest known written European references to the durian were in the travel records of Niccolò da Conti in the 15th century. The trees migrated to the Americas, Australia, and Africa in the twentieth century.

TREE CHARACTERISTICS. A tropical evergreen tree, under ideal conditions mature trees can grow up to 150 feet tall with straight tree trunks reaching almost four feet in diameter. (But trees grown in orchards and backyards are considerably smaller. Grafted trees tend to grow around 32 feet tall.) The tree bark is dark reddish brown. The tree branches are almost horizontal. The oval shaped leaves grow up to about eight inches long and three inches wide. The top of the leaves and shiny green, and the underside is somewhat hairy and reddish brown.

The tree bears fragrant blossoms with nectar that grow in clusters. The clusters can have up to 25 blossoms. Each blossom is about two to two and two-thirds inches long and has five petals. The blossoms grow on the underside of mature branches. The blossoms are the color of the pulp that will grow inside the fruit. For example yellow blossom will produce durian with yellow pulp. White blossoms will produce durian with white pulp, etc. The blossoms open in the evening and are pollinated by bats, moths, bees and other night flying pollinators. The blossoms can also be hand pollinated.

Trees are propagated by seed or by grafting. Trees grown from seed take eight to 15 years to begin bearing fruit. Grafted trees take only four to eight years to begin bearing fruit. The trees generally live 80 to 150 years. Older trees produce less fruit, but the fruit quality is said to be better. There are more than 200 cultivars of durian. An odorless cultivar, 'Chantaburi No. 1' also exists.

▶ FAMILY: Bombacaceae. GENUS: *Durio* Adanson. SPECIES: *Durio zebethinus* Murray.

SELECTING A TREE. Two major considerations in selecting a durian tree are how much space you have for it, and how soon you want it to bear fruit. If you have the patience to wait until a tree begins to fruit, and have lots of space to grow it, then you may find a tree grown from a seedling suitable.

However, if you prefer a tree that begins to bear fruit in half the time as one grown from a seedling, and doesn't grow as tall, then you'll want a grafted tree.

When selecting a tree from the nursery, check the tree's overall health. Make sure its leaves are not discolored, and free from insects or signs of insect damage. Check the entire tree to make sure it is free from injury, such as broken branches, cuts in the trunk, etc. Check to make sure the tree trunk is not sunburned. Also visually inspect the pot soil to make sure there is no fungi growing from the soil, indicating uneven or over watering that can affect the overall health of the trees.

GROWING A DURIAN TREE. Durian trees are considered ultra-tropical trees so they require hot, humid temperatures. They thrive in areas where the average temperature is a minimum of 72 degrees Fahrenheit. They will not withstand any frost or a drought. If you live on the mainland in the United States, it will be very challenging growing a durian tree without a tropical greenhouse. The trees grow best in rich, well draining loamy soil, such as a sandy clay, or clay loam. The well draining soil is important to prevent root rot from developing. They prefer soil with a pH level of 5 to 6.5.

To plant a tree, dig a hole about two feet deep. Place the seedling in the hole so that the root crown is at land level. Fill back in the hole with the original soil mixed with compost. Water thoroughly and frequently to insure the seedling is not stressed out by a lack of water. Newly planted seedlings require 50 percent shade until they have a little over a yard of new growth. After that you can slowly increase the amount of sun they get over a year until they are at full sun.

The trees respond well to the application of manure and a balanced fertilizer. The rate of fertilization should be increased as the tree ages. Pruning should be done in young trees to better shape the tree. Prune away any excess trunks since there should be only one. And since fruit forms on horizontal branches, upright branches should be pruned. The major disease of durian trees is *Phytophthora*, a pathogen which prevents the tree from absorbing the water and nutrients it needs. It frequently results in root rot.

▶ USDA Hardiness Zone: 11. CHILLING HOURS: 0. WATER REQUIREMENTS: 60–150 inches of rainfall annually.

HOW TO CONSUME. To eat a fresh durian, place it on a cutting board. You may want to use a kitchen towel or gloves to protect your hand while you hold the durian in place on the cutting board. Use a knife to cut through the hard outer skin and make a slice through the skin completely around the fruit. Next, carefully pull the skin and pulp apart so that you have two halves of the fruit. Peel out the segments of fruit with your fingers, or a spoon. Remove the large seeds from the pulp, and enjoy eating the durian.

The durian seeds are also edible but they need to be cooked prior to consumption to destroy the toxic cyclopropene fatty acids present in the raw seeds. A common method of cooking them is to boil them in water, with salt if desired, like you would boil a potato. Peel away the seed skin and consume them. They are said to taste like a mild potato. Other methods of cooking the seeds include roasting them, and slicing and frying them in oil.

Fresh durians can also be frozen. Remove the durian pulp from the fruit, place it in a plastic freezer bag or container, and put it in your freezer. Be sure your freezer temperature is set to zero degrees Fahrenheit or lower. Durian pulp can also be dehydrated. Durians are frequently used in baked goods such as cakes, cookies, pies and pastries. In Asian countries you'll also find durian ice cream. Commercially, durians are also canned, frozen, and made into and sold as chips, candies, and other sweet treats.

ADDITIONAL INFORMATION. WARNING: Durians are high in natural sugars. Eating significant amounts of durians elevates blood glucose levels. Diabetics should either use caution in consuming durians, or avoid them altogether.

Durians are high in vitamin C. They also contain minerals like potassium, magnesium, and manganese, and the amino acid tryptophan.

Why do durians smell so bad? A study published in *Nature Genetics* reported that researchers identified a class of genes that regulate the production of odor compounds called volatile sulphur compounds (VSC). Their research showed that VSC production in durians was supercharged, resulting in their strong odors. The durian's strong smell attracts animals to consume them. The durian inspired J. Corner's "Durian Theory" of the evolution of tropical forest plants. The theory illustrates the interdependence of plants and animals, the plants providing food like durians for animals like elephants, and the animals consuming the fruit and providing dispersal of the undigested seeds.

Most of the durians you'll find on the mainland U.S. will be in the frozen food section of your local Asian market.

REFERENCES

"About Durian." Northern Territory Government, Australia website. https://nt.gov.au/environment/home-gardens/growing-vegetables-at-home/durian. Accessed 05/18/18.

Crabb, Lindy. "Fruits of Paradise." *Good Health (Australia Edition)*, January 2017, p. 70–71.

David, Cynthia. "Durian Fruit Smells Like Hell and Tastes Like Heaven." *The Hamilton Spectator*, August 22, 2007, Food Section, p. G5.

"Durian." *Chatelaine*, vol. 81, no. 8, August 2008, p. 28.

"Durian." Datasheet. Centre for Agriculture and Biosciences International. January 2018. https://www.cabi.org/isc/datasheet/20179. Accessed 05/12/18.

Durian (Durio zebethinus). National Agricultural Research Institute, Papua New Guinea, 2007.

Durian Production Guide. Phillipine Department of Agriculture, Bureau of Plant Industry, Davao National Crop Research, Development & Production Support Center. 2017.

Durian—The King of Tropical Fruits. Homegrown Biotech, Kerala, India. n.d.

"Durians—Assorted." Tropical Fruit Farm, Malaysia. http://www.tropicalfruits.com.my/pdf/Durians-Assorted-k.pdf. Accessed 05/13/18.

"*Durio zebethinus* (Bombacacea)." Montoso Gardens website. http://www.montosogardens.com/durio_zibethinus.htm. Accessed 05/12/18.

"*Durio zebethinus—L.*" Plants for a Future. www.pfaf.org/user/Plant.aspx?LatinName=Durio+zibethinus. Accessed 05/20/18.

"*Durio zibethinus* Murray." USDA, NRCS. 2018. The PLANTS Database. National Plant Data Team, Greensboro, NC 27401-4901 USA. https://plants.sc.egov.usda.gov/core/profile?symbol=GAMA10. Accessed 12/10/18.

"The Fruits of Winter." *Newsweek,* vol. 151, no. 4, January 28, 2009, p. 61.

Genthe, Henry. "Durians Smell Awful—But the Taste, Say the Brave, Is Heavenly." *Smithsonian,* vol. 30, no. 6, September 1999, p. 94–102.

"How to Grow Durian." www.durianhaven.com/how-to-grow-durian.html. Accessed 05/13/18.

"How-to Guide to Growing Durian: A General Reference." Cropsreview.com. www.cropsreview.com/durian.html. Accessed 05/13/18.

Jenkins, Mark. "Ah, Sweet Durian." *Washington Post,* March 25, 1990. www.washingtonpost.com/archive/lifestyle/travel/1990/03/25/ah-sweet-durian/bd67d84f-7790-4a3f-a46c-8d1806623f24/?utm_term=.d397103bfba7. Accessed 05/13/18.

Kingsbury, Kathleen. "Songpol Somsri." *Time,* vol. 170, no. 14, October 1, 2007, p. 69.

Masri, M. "Flowering, Fruit Set and Fruitlet Drop of Durian (*Durio zebithinus* Murr.) Under Different Soil Moisture Regimes." *Journal of Tropical Agriculture and Food Science,* vol. 27, no. 1, 1999, p. 9–16.

Morton, Julia F. *Fruits of Warm Climates.* Eugene, OR: Wipf and Stock Publishers, 2003.

Nagy, Steven, and Philip E. Shaw. *Tropical and Subtropical Fruits: Composition, Properties and Uses.* Westport, CT: AVI Publishing. 1980.

Orwa, C., et al. "*Durio zebethinus.*" 2009 Agroforestree Database: A Tree Reference and Selection Guide Version 4.0. www.worldagroforestry.org/treedb/AFTPDFS/Durio_zibethinus.PDF. Accessed 05/12/18.

Paull, Robert E., and Saichol Ketsa. "Durian: Postharvest Quality—Maintenance Guidelines." Fruit, Nut and Beverage Crops F_N-27. College of Tropical Agriculture and Human Resources, University of Hawaii at Manoa, 2014. www.ctahr.hawaii.edu/oc/freepubs/pdf/F_N-27.pdf. Accessed 05/10/18.

Piper, Jacqueline. *Fruits of South-East Asia: Facts and Folklore.* Oxford, NY: Oxford University Press, 1989.

"Revealed: Secret Behind Durian Fruit's Pungent Smell." DNA website. www.dnaindia.com/science/report-revealed-secret-behind-durian-fruit-s-pungent-smell-2551798. Accessed 05/16/18.

Shenon, Philip. "Love It or Hate It, This Is the Forbidding Fruit." *New York Times,* July 18, 1994. www.nytimes.com/1994/07/18/world/singapore-journal-love-it-or-hate-it-this-is-the-forbidding-fruit.html?mcubz=1. Accessed 05/13/18.

Stromberg, Joseph. "Why Does Durian Fruit Smell So Terrible?" *Smithsonian Magazine.* November 30, 2012. www.smithsonianmag.com/science-nature/why-does-the-durian-fruit-smell-so-terrible-149205532/. Accessed 08/16/17.

Stuppy, Wolfgang. "Durian—The King of Fruit." Kew Royal Botanic Garden. www.kew.org/blogs/archived-blogs/durian-king-fruit. Accessed 05/13/18.

Teh, Bin Tean, et al. "The Draft Genome of Tropical Fruit Durian (*Durio zebithinus*)." *Nature Genetics,* vol. 49, no. 11, November 2017, p. 1633–1641.

"Today's Special … Odourless Durians." *Maclean's,* vol. 120, no. 16, April 30, 2007, p. 64.

Walsh, Robb. "The Fruit I Can't Get Past My Nose." *Natural History,* vol. 108, no. 7, September 1999, p. 76–77.

Watson, B.J. "Durian." Fact Sheet #8. Cairns, QLD: Rare Fruit Council of Australia. 1983.

Wayo, Kanuengnit, et al. "Bees Are Supplementary Pollinators of Self-Compatible Chiropterophilous Durian." *Journal of Tropical Ecology,* vol. 34, no. 1, January 2018, p. 41–52.

Fig   **70**

Wedner, Diane. "Far-Out Foods." *National Geographic Explorer*, vol. 10, no. 3, November/December 2010, p. 10–15.
"Worse Than a Stinking Hangover." *New Scientist*, vol. 203, no. 2725, September 12, 2009, p. 4.
Year of the Durian website. www.yearofthedurian.com. Accessed 05/12/18.

Fig (*Ficus carica* L.)

The fig, also known as the common fig and edible fig, is a "false fruit," technically not a fruit. Botanically it's a syconium, i.e., a fleshy receptacle containing numerous flowers. But for the sake of this section, it will be referred to as a fruit. Somewhat pear shaped, the fig generally measures from about one to two inches long, and has a somewhat thin, tender, light green skin that can turn brown or purplish brown as it ripens, depending upon the cultivar. The inner wall of the skin is fleshy and whitish yellow. Ripe fig pulp is sweet and juicy and ranges in color from pink to a purplish pink depending upon the cultivar. Most of the pulp contains a mass of small edible seeds. Fresh figs taste like honey with a hint of berries.

FIG LORE. There are many myths regarding the origin of the fig tree. For example in one, the Titan Lyceus, who was being pursued by Jupiter, was turned into a fig tree by the goddess Rhea. In another myth, the fig tree is the offspring of Oxylus, the King of Ellis, with a Hamadryad. In yet another myth, Bacchus, the Roman god of agriculture, is said to have introduced the sacred tree to humans, which is why he is often depicted as wearing a laurel wreath of fig leaves.

In Hindu and Buddhist mythology the fig tree is considered sacred because it's believed that it was the tree under which Siddhartha Gautama (Buddha) found enlightenment in 528 BCE. The fig is often believed to have been the first cultivated fruit tasted by man: "beneath the boughs of the figtree Adam hid himself after having eaten the forbidden fruit; with its leaves he endeavoured to hide his nakedness."

TREE HISTORY. Originating in western Asia, the fig dates back to ancient times. It's believed that humans spread it throughout the Mediterranean. Remnants of figs were found in archeological sites dating back to at least 5,000 BCE. Figs were introduced to England in the early 1500s, into Mexico, South Africa and Australia in the mid–1500s, and into the United States in the 1660s.

TREE CHARACTERISTICS. The fig tree is a deciduous tree growing about 10 to 30 feet tall with a rounded crown. Its trunk is seldom larger than seven inches wide. The bark is silvery gray. It has a shallow and spreading root sys-

Fig *(Couleur/pixabay.com)*

tem. Fig trees feature alternate simple dark green leaves. The leaves usually measure four to eight inches long and are generally five-lobed. The tree bears small greenish blossoms twice in the season. Common fig blossoms do not need pollination to set a crop. But it should be noted that Smyrna and Caprifig figs, also known as European figs, require pollination by caprifig wasps generally not found in the U.S. These varieties are usually not available to the home grower in the U.S.

Most fig trees require a considerable period of heat for the fig to ripen. But cold hardy fig cultivars are available. Fig trees can be espaliered and common figs reportedly can be grown indoors. The tree and unripe figs contain a milky latex. The latex can cause skin irritation in sensitive people. There are a number of common fig cultivars. Some cultivars are better adapted for hot dry climates or cool coastal climates than others. For example, the 'Alma' and the 'Panachee' require long warm summers, and the 'Conadria' does best in hot climates. Whereas the 'King,' 'Osborne Prolific' and 'White Genoa' cultivars do well in cooler areas.

Popular cultivars include the following: 'Alma'—figs have light yellow skin and yellowish tan pulp. • 'Black Mission'—figs have purple black skin and red pulp. • 'Brown Turkey'—bears large figs with purplish green skin and red pulp. • 'Conadria'—greenish yellow skin with a purple blush and strawberry colored pulp. • 'Excell'—produces large yellow skinned figs with yellowish

Fig   72

pulp. • 'Green Ischia'—small greenish yellow skinned figs with strawberry colored pulp. • 'Kadota'—yellowish green skinned figs with amber pulp. • 'King'—large figs with dark green skin and purple pulp. • 'Osborne Prolific'— fig have a dark reddish brown skin and amber pulp. • 'Panachee'—greenish yellow skin with dark green strips and strawberry colored pulp. • 'Texas Ever-bearing'—figs have mahogany-purple skin and strawberry colored pulp. • 'White Genoa'—large yellow green skinned figs with strawberry colored pulp. • 'Black Jack'—dwarf tree, purple skinned figs with strawberry colored pulp.

Fig trees can be propagated by seed, by cuttings of two to three year old wood, and by air layering. Fig trees generally do not start to fruit until they are two to six years of age.

▶ FAMILY: Moraceae. GENUS: *Ficus* L. SPECIES: *Ficus carica* L.

SELECTING A TREE. Select a cultivar that is best suited for your climate. Also, take into consideration that some cultivars produce figs that are better dried than eaten fresh, and vice versa.

When selecting a tree from the nursery, check the tree's overall health. Look for active growth, such as new or young leaves. Make sure its leaves are

Dried figs *(Chefallen/Wikimedia Commons)*

not yellow or discolored, and free from insects or signs of insect damage. Check the entire tree to make sure it is free from injury, such as broken branches, cuts in the trunk, etc. And if possible, gently lift the tree out of the pot to check the roots. Overgrown trees that have seriously girdled roots will likely develop a poor root structure so they should be avoided.

How to Grow a Fig Tree. Fig trees prefer full sun, and grow in a wide range of soils ranging from light sand to clay soils, provided there is adequate drainage. High acid soils should be avoided. It prefers soils with a pH between 6.0 and 6.5. To plant a tree, select a sunny location that can accommodate the size of a mature tree. Dig a hole at least twice the diameter of the pot the tree came in and about as deep as the container. Place the tree in the hole, loosening or cutting away any girdled roots. The tree should be planted so that the top of the root ball is about one to two inches higher than the soil level. Refill the hole with the original soil. Water the newly planted tree thoroughly. The tree should be regularly watered until it becomes established. Fig tree pests include nematodes, beetles and flies.

▶ USDA Hardiness Zones: 8–11. Chilling Hours: 100. Water Requirements: Average.

How to Consume. Ripe figs are most commonly enjoyed eaten fresh. Ripe figs should be soft to your touch. To eat them fresh, you can either peel the skin back from the stem and consume the pulp without utensils, or you can grab a knife and slice the fig in half and scope out the pulp with a spoon. Fresh figs have a short shelf life at room temperature. But they can be stored in the refrigerator for about a week. Fresh ripe figs can also be frozen. Simply wash them, and peel them if desired. Lay them out single layer on tray in the freezer at zero degrees Fahrenheit or lower. When frozen thoroughly, simple pack them in an airtight freezer container and place them back in your freezer. The figs can also be frozen in a 40 percent syrup pack.

Figs are frequently dried. To dry them, thoroughly wash them. Slice them in half and dry them in your electric dehydrator for six to 12 hours until they are dried to about 20 percent moisture. To test for dryness, cut a couple of pieces in half. You should not be able to squeeze any moisture out of them, and there should be no visible moisture. And if you fold a piece over, it should not stick to itself. Once the fruit is dry, let it cool for an hour. Then package the fruit preferably in a freezer container or glass jar. Freezer bags can also be used but unfortunately they are not rodent proof. Store the container in a dark, dry, cool place for no more than a year.

Figs can be used in baked goods such as breads, cookies, puddings, pies, etc. Figs can also be canned, made into jam, pickles and other preserves. For instructions on preserving figs in these and other methods, and for safe recipes, visit the National Center for Home Food Preservation at www.nchfp.uga.edu.

ADDITIONAL INFORMATION. In India fig leaves are used as fodder. In Latin American folk medicine, figs are used with other substances to treat swollen gums and sore throats.

REFERENCES

Andersen, Peter C., and Timothy E. Crocker. "The Fig." HS27. University of Florida, Institute of Food and Agricultural Sciences Extension. June 2016. https://edis.ifas.ufl.edu/mg214. Accessed 12/23/18.
"*Ficus carica.*" Oregon State University, College of Agricultural Sciences, Department of Horticulture. https://landscapeplants.oregonstate.edu/plants/ficus-carica. Accessed 12/23/18.
"*Ficus carica* L." USDA, NRCS. 2018. The PLANTS Database. National Plant Data Team, Greensboro, NC 27401-4901. https://plants.sc.egov.usda.gov/core/profile?symbol=FICA. Accessed 12/23/18.
"Fig." California Rare Fruit Growers. www.crfg.org/pubs/ff/fig.html. Accessed 12/23/18.
Folkhard, Richard. *Plant Lore, Legends and Lyrics: Embracing the Myths, Traditions, Superstitions, and Folk-Lore of the Plant Kingdom.* London: Sampson Low, Marston, Searle, and Rivington. 1884.
Hill, Lewis, and Leonard Perry. *The Fruit Gardener's Bible.* North Adams, MA: Storey Publishing, 2011.
Ingels, Chuck, et al., editors. *The Home Orchard: Growing Your Own Deciduous Fruit and Nut Trees.* Richmond, CA: University of California, Agriculture and Natural Resources, 2007.
Morton, Julia F. *Fruits of Warm Climates.* Eugene, OR: Wipf and Stock Publishers, 2003.
Nagy, Steven, and Philip E. Shaw. *Tropical and Subtropical Fruits: Composition, Properties and Uses.* Westport, CT: AVI Publishing. 1980.
Ortho All About Citrus and Subtropical Fruits. Des Moines, IA: Meredith Books, 2008.
"Scientific Name: *Ficus carica.*" Arizona State University. www.public.asu.edu/~camartin/plants/Plant%20html%20files/ficuscarica.html. Accessed 12/23/18.
Stover, Ed, et al. "The Fig: Overview of an Ancient City." *HortScience*, vol. 42, no. 5, August 2007, p. 1083–1087. https://pubag.nal.usda.gov/download/7459/PDF. Accessed 12/23/18.

Ginkgo (*Ginkgo biloba* L.)

The ginkgo fruit is round to oval shaped, measuring one to three inches in length, about the size and shape of an apricot. Its skin is green, turning yellow when mature. The fruit emits a foul odor often compared to vomit. The fruit's pulp is not edible, and contains urushiol, the chemical in poison ivy.

The seed in the center of ginkgo is similar in size and shape to an olive. The seed is often referred to as a nut because of its appearance. Raw seeds should never be consumed. However, they can be eaten when cooked. The seeds are toxic, but can be consumed in small quantities. More data on the toxicity of the ginkgo pulp and seeds appears in the Additional Information section. Technically what's referred to as the ginkgo fruit is a seed with a fleshy outer covering.

Ginkgo *(www.needpix.com)*

GINKGO LORE. There's a tale about the Jotakuji Temple, better known as the Ginkgo Temple, in Japan. In 1274, Nichiren shonen, a Buddhist priest, took on a challenge to discuss the teachings of Buddhism with another Buddhist priest named Echo. Echo lost, causing him to develop a grudge against Nichiren shonin. So he conspired with the head priest at the Jotakuji temple to kill Nichiren shonin with poisoned rice cakes. Echo offered the rice cakes to him when a white dog suddenly appeared. Being kind-hearted, Nichiren shonin saw the dog and fed him the rice cakes from Echo. As soon as the dog ate them, it fell to the ground in agony. Realizing Nichiren shonin must be powerful, Echo confessed. He apologized and asked to become a disciple of Nichiren shonin again. Nichiren shonin gave the dog an antidote and he lived many more years. The white dog was believed to be an incarnation of various gods who were sent there to protect Nichiren shonin. When the dog died, it was buried and Nichiren shonin put his stick, made of ginkgo, on top of the burial mound as a marker. The stick began to grow, eventually becoming a huge tree. The seeds are borne on the surface of the leaves, and seeds are fang shaped, just like the teeth of the white dog.

Through the ages, ginkgo and ginkgo extracts have been said to cure everything from cancer to multiple sclerosis, but there's a lack of scientific evidence to support these many uses. Some initial studies have indicated gingko may improve symptoms of Alzheimer's and other mental conditions, but significantly more research needs to be done.

TREE HISTORY. The ginkgo tree is one of the oldest trees on the planet, dating back 200 million years, to the time of the dinosaurs. Often considered a "living fossil," fossil records indicate the tree appears unchanged through the ages. What also makes the tree unique is that it isn't closely related to any other living plant.

Originating in China, it made its way to Japan, and from there was introduced into Europe in about 1730. It officially made its way to North America, specifically Philadelphia in 1784. The tree is currently documented as growing in Asia, Turkey, the Czech Republic, Slovakia, Estonia, Germany, Russia, South America and the United States.

TREE CHARACTERISTICS. Also known as a maidenhair tree, because the leaves resemble some of the pinna (leaflets) of the Maidenhair fern, ginkgo trees can reach heights from 40 to more than 100 feet tall, and up to 60 feet wide. However, a number of cultivars have been bred to serve as bonsai and other small trees. Ginkgo trees are long living trees, with some claimed to be over 2500 years old. The trees are disease and insect resistant, as well as drought tolerant. The tree species is very adaptable. It tolerates air pollution and wind, and grows in many soil types and pH levels. They are very resilient trees. Four ginkgo trees even survived the atomic blasts in Hiroshima, Japan.

A deciduous tree, it has unique fan shaped glossy green leaves approximately two to four inches long, usually with two lobes between the leaf vein. The leaves turn bright yellow in the autumn and seem to all fall off the tree suddenly at the same time in the fall. Older tree branches also develop unique distinct gray spurs. The tree's blossoms are green, and have a pleasant fragrance. The tree's trunk is smooth and gray when young, but in mature trees the trunks have deep brown furrows.

Trees are either male or female. Male trees do not produce fruit and are generally preferred as ornamental trees. Male trees have cone-like spiral clusters of pollen sacs on short shoots. The shoots have three to six pollen sacs. Female trees produce ovules (two) at the ends of stalks. Female trees produce ginkgo fruit when the ovules are pollinated by male trees. Many find the aroma of the ginkgo fruit very offensive. Trees do not begin to produce pollen sacs or ovules (depending upon the sex of the tree) until they are 20 to 25 years of age.

If you are growing a tree from seed, you won't know if you've grown a male or female tree for 20 to 35 years. If you only want to grow a male tree, consider purchasing a tree propagated by male cuttings grafted onto seedling root stock. If you have a female tree because you want it to bear fruit, but don't have space to plant a male tree, you'll need to graft male branches onto the female tree for pollination.

▶ FAMILY: Ginkgoaceae. GENUS: *Ginkgo* L. SPECIES: *Ginkgo biloba* L.

SELECTING A TREE. When selecting trees, remember there are both male and female trees. If you want trees that produce ginkgos, you'll need both a male and female tree, or a female tree with grafted male branches on the tree for pollination. Be wary of trees propagated by seed because you will not know their sex for at minimum 20 years. If you want a ginkgo tree as an ornamental tree, select a male tree.

There are a variety of cultivars available. Popular male cultivars include the following: 'Autumn Gold'—broad pyramidal shape, it's popular for the striking golden yellow color its leaves turn in the fall. • 'Magyar'—upright, narrow pyramidal form. • 'Princeton Sentry'—the tree is a fastigiate, meaning it has upright branches that taper at the top. • 'Saratoga'—has an upright habit and narrow foliage. • 'Jade Butterflies'—popular as a slow growing dwarf tree. The name is based upon the trees leaves that appear in pairs in symmetry so they resemble a butterfly. • 'Mariken'—another dwarf cultivar, has low compact branches.

Popular female cultivars include: 'Variegata'—has variegated green and yellow leaves, is shrub like in form. • 'King of Dongting Mountain'—slow growing cultivar with large leaves.

HOW TO GROW A GINKGO TREE. Ginkgo trees grow in a wide variety of soils, but prefer deep, moist, well draining soil. It grows in soils with a pH from 5.0 to 8.2. The trees prefer full sun.

To plant a tree, dig a hole at least twice the diameter of the pot the tree came in, about as deep as the container. Place the tree in the hole, loosening or cutting away any girdled roots. The tree should be planted so that the top of the root ball is about one to two inches higher than the soil level. Plant it so that the root flare is visible. Refill the hole with the original soil. Water the newly planted tree thoroughly. Regularly water the tree, keeping the soil moist until the tree becomes established.

Prune young ginkgo trees so they develop a strong central leader. The trees are fairly disease and pest resistant.

HOW TO GROW FROM SEED. If you plan to grow a tree from seed, just bear in mind that you won't be able to determine if your seedling is a male or female tree until it matures enough to bear either pollen sacs or ovules starting at 20 to 35 years of age. When gathering fresh dropped ginkgos to plant, be sure to wear plastic or latex gloves to protect your hands. The outer covering/pulp of the ginkgo contains urushiol, the same irritant found in poison ivy. So you don't want the ginkgo to touch your skin.

Thoroughly clean away any flesh from the seed. Wash the seeds in clean water and dry them. Cold seed stratification, although not required, is suggested by some sources to improve germination rates. (Cold seed stratification is when you subject a seed to a cold and moist environment for a period of time.) Place the seeds between moist paper towels, seal in a plastic bag, and place in your refrigerator for two months.

The seeds need to be scarified before planting. To mechanically scarify them, use a sharp knife to nick the seed or rub sandpaper against the seeds to weaken the seed coat and allow moisture to penetrate easier. Next soak the seeds in warm water for 24 hours. Once the 24 hours are up, place the seeds in a solution of 10 percent bleach and 90 percent water for 10 minutes. This will help protect the seeds from fungus. Rinse the seeds thoroughly. Plant the seeds in a well draining planting medium. Keep the soil moist. The seeds will germinate within 60 days.

▶ USDA HARDINESS ZONES: 3–8. CHILLING HOURS: 0. WATER REQUIRE-MENTS: Trees require regular watering after planting until established. Mature trees only require occasional watering.

HOW TO CONSUME. Because ginkgo seeds contain cyanogenic glycosides, they need to be cooked before being consumed. They can be boiled or baked, or roasted. To cook them, clean away any pulp around the seed. Be sure to wear plastic or latex gloves when touching any ginkgo pulp. Wash the seeds in water and pat dry.

You can roast them like you would any other raw nut. Likewise you can also microwave them like you would any other raw nut as well. To bake them, lay them out on a baking pan and bake at 300 degrees Fahrenheit for 30 minutes to an hour. However you decide to cook them, be sure to remove the paper like seed skin before consuming.

Ginkgo seeds are sometimes added to soups and other dishes in Southeast Asia. Because ginkgo nuts contain a toxin, experts suggest only a very small number should be consumed daily. Children should not consume ginkgo seeds.

ADDITIONAL INFORMATION. TOXICITY WARNING: Raw ginkgo seeds contain the toxin cyanogenic glycosides and should never be consumed. Ginkgo seeds also contain the compound 4-methoxypyridoxine (also known as ginkgotoxin, a neurotoxin) which is not destroyed by heat. Therefore ginkgo seeds should not be consumed by children, and adults should only eat a few seeds. Some signs of ginkgotoxin poisoning can include nausea, vomiting, difficulty breathing, seizures and convulsions. The ginkgo flesh and pulp contain the toxins alkyl phenol and urushiol and should never be consumed. Urushiol is the same toxin found in poison ivy; hence you should always wear gloves when handling ginkgos. This will prevent skin irritation.

There are a number of major and minor drug interactions with ginkgo listed on the WebMd website. For major interactions, the site advises not taking ginkgo with ibuprofen, anticoagulant and antiplatelet drugs. Because ginkgo can slow blood clotting, it could increase the risk of bleeding and bruising. Medical professionals also recommend not taking ginkgo or ginkgo supplements if you are taking oxidase inhibitors (MAOI) or anti-depressants, or if you are pregnant. If you are, taking ginkgo can increase your risk of gastrointestinal problems, diarrhea, nausea, vomiting, dizziness and bleeding.

The name Ginkgo stems from the Chinese and Japanese word "ginkyo" which translates into "silver apricot."

REFERENCES

Baird, Sarah. "Ginkgo Nuts: What the Hell Are They, and Why Aren't We Eating Them?" *GQ*, April 26, 2017. www.gq.com/story/ginkgo-nuts-snack-cooking. Accessed 09/22/18.
Chase, Jeri. "The Ginkgo—A True Living Fossil." *Forests for Oregon*, Summer 2007, p. 17–19.
"Choosing the Right Tree for the Right Place: Ginkgo Biloba." South Carolina Urban and Community Forestry Council. www.clemson.edu/cafls/vincent/articles/ginko_bioba.pdf.
Clark, Patterson. "Harvesting Ginkgo Fruits: Breaking the Stink Barrier." *Washington Post*, October 12, 2010. http://www.washingtonpost.com/wp-srv/special/metro/urban-jungle/pages/101012.html?noredirect=on. Accessed 07/17/18.
Coder, Kim D. "Ginkgo." Pub. No. 22. Warnell School of Forestry and Natural Resources, University of Georgia, April 2016. www.warnell.uga.edu/sites/default/files/publications/WSFNR-16-22%20Coder.pdf. Accessed 07/16/18.
Cohn, Roger. "Interview—Ginkgo: The Life Story of the Oldest Tree on Earth." *Yale Environment 360* (Yale School of Forestry and Environmental Studies), May 1, 2013. https://e360.yale.edu/features/peter_crane_history_of_ginkgo_earths_oldest_tree. Accessed 07/17/18.
Del Tredici, Peter. "The Ginkgo in America." *Arnoldia*, vol. 41, no. 4, July/August 1981, p. 150–161. http://arnoldia.arboretum.harvard.edu/pdf/articles/1981-41-4-the-ginkgo-in-america.pdf. Accessed 09/22/18.
Diamond, B.J., and M.R. Bailey. "Ginkgo Biloba: Indications, Mechanisms, and Safety." *Psychiatric Clinics of North America*, vol. 36, no. 1, March 2013, p. 73–83.
Easton, John. "Ginkgo Rhymes with Stinko." *ScienceLife* (University of Chicago Hospitals), November 13, 2015. https://sciencelife.uchospitals.edu/2015/11/13/ginkgo-rhymes-with-stinko/. Accessed 07/17/18.
Eaton, Joe, and Ron Sullivan. "Ginkgo Biloba Trees Show Their Colors." SF Gate. December 26, 2010. www.sfgate.com/homeandgarden/thedirt/article/Ginkgo-biloba-trees-show-their-colors-2452081.php. Accessed 07/17/18.
Fitzpatrick-Cooper, Julia. "Plant Profiles: HORT 2242 Landscape Plants II. Botanical Name: *Ginkgo biloba*." College of Du Page. 2014. https://dc.cod.edu/horticulture-2242-ginkgo-biloba/ginkgo-biloba.pdf. Accessed 09/21/18.
Gilman, Edward F., and Dennis G. Watson. *Ginkgo Biloba*. Fact Sheet ST-273. U.S. Dept. of Agriculture, U.S. Forest Service, November 1993.
"Ginkgo." *Horticulture Week*, January 28, 2016. www.hortweek.com/ginkgo/ornamentals/article/1381260. Accessed 09/26/18.
"Ginkgo." WebMd. www.webmd.com/vitamins/ai/ingredientmono-333/ginkgo. Accessed 07/17/18.
"Ginkgo biloba." North Carolina State University, Cooperative Extension Services. https://plants.ces.ncsu.edu/plants/all/ginkgo-biloba/. Accessed 07/17/18.
"*Ginkgo biloba*—L." Plants for a Future. https://pfaf.org/user/plant.aspx?LatinName=Ginkgo+biloba. Accessed 07/17/18.
"*Ginkgo biloba* L." USDA, NRCS. 2018. The PLANTS Database. National Plant Data Team,

Greensboro, NC 27401-4901. https://plants.sc.egov.usda.gov/core/profile?symbol=GIBI2. Accessed 12/26/18.

"Ginkgo *Ginkgo biloba*." Arbor Day Foundation. www.arborday.org/trees/treeguide/Tree Detail.cfm?itemID=1092. Accessed 07/18/18.

"Ginkgo (*Ginkgo biloba*)" Nebraska Forest Service. https://nfs.unl.edu/CommunityForestry/ Trees/Ginkgo.pdf. Accessed 07/16/18.

"The Ginkgo Pages." https://kwanten.home.xs4all.nl/index.htm. Accessed 07/16/18.

"Ginkgoaceae." Iowa State University, Department of Ecology, Evolution, Organismal Biology. www.eeob.iastate.edu/classes/bio366/families/Ginkgoaceae.pdf. Accessed

"Ginkgoaceae: Ginkgo Family." Princeton University Press. http://assets.press.princeton.edu/ chapters/s10217.pdf.

"How to Safely Harvest and Prepare Ginko Nuts." Instructables.com. www.instructables.com/ id/How-to-Safely-Harvest-and-Prepare-Ginko-Nuts/. Accessed 07/17/18.

Kajiyama, Y., et al. "Ginkgo Seed Poisoning." *Pediatrics*, vol. 109, no. 2, February 2002, p. 325–327.

"Maidenhair Tree, Ginkgo—Ginkgo biloba." University of California, Agriculture and Natural Resources, Integrated Pest Management. http://ipm.ucanr.edu/PMG/GARDEN/PLANTS/ maidenhair.html. Accessed 07/16/18.

Paulson, John. *Propagation of Ginkgos*. North Carolina State University. www.ndsu.edu/pub web/chiwonlee/plsc368/student/papers02/jpaulson/ginkgopropagation.html. Accessed 07/17/18.

Pittenger, Dennis. *California Master Gardener Handbook*. 2nd ed. University of California, Agricultural and Natural Resources. 2014.

"Plant Guide: Ginkgo." U.S. Dept. of Agriculture, Natural Resources Conservation Service. https://plants.usda.gov/plantguide/pdf/pg_gingk.pdf. Accessed 07/16/18.

Shepperd, Wayne D. *The Woody Plant Seed Manual. Agriculture Handbook No. 727*. Washington, D.C. U.S. Dept. of Agriculture, Forest Service. 2008. www.fs.fed.us/rm/pubs_ other/wo_AgricHandbook727/wo_AgricHandbook727_559_561.pdf. Accessed 09/22/18.

Sierpina, V.S., et al. "Ginkgo biloba." *American Family Physician,* vol. 68, no. 5, September 2003, p. 923–926. www.ncbi.nlm.nih.gov/pubmed/13678141. Accessed 09/29/18.

"Smart Tree Selections for Communities and Land Owners: Ginkgo." Michigan State University Extension, Dept. of Forestry and Horticulture. http://www.canr.msu.edu/home_ gardening/uploads/files/SmartTreeSelections/Ginkgo-AltTrees.pdf. Accessed 07/16/18.

"A Tree of Many Stories." Rutgers Gardens, Rutgers School of Environmental and Biological Sciences. Rutgers, The State University of New Jersey. October 2016. http://rutgersgardens. rutgers.edu/pdfs/Plant%20of%20the%20Month/Plant%20of%20the%20Month,%202016/ 10)%20POM%20October%202016%20Gingko.pdf. Accessed 07/16/18.

Goji Berry (*Lycium barbarum and Lycium chinesis*)

The berries from two different species of the botanic genus *Lycium L.* are marketed as goji berries. Besides being labeled as goji berries, *Lycium* barbarum, also known as Chinese Matrimony Vine and Chinese Wolfberry, the Duke of Argyll's tea, and *Lycium chinesis* also known as Chinese Boxthorn and Chinese Desert Thorn, both produce berries. Goji berries range in color from orange to deep red, are oblong, and measure up to three-quarters of an inch long. They have tiny seeds.

Fresh goji berries are somewhat tart. Their taste is often described as being unique, but likened to a cross between a sour cherry and a sour plum. Dried goji berries are frequently described as tasting somewhat similar to a dried cranberry, but not as sweet and with a slight bitter aftertaste.

GOJI BERRY LORE. Myths and legends about the goji berry abound. One common myth dates back to the era of the first emperor of China (about 2800 BCE). The story tells of the people of the Hunza in the Himalayas who all live extraordinarily long lives, well into their hundreds. Their secret is said to be the goji berries growing near and falling into their water supply, thereby creating a fountain of youth.

Another story dates back to the Tang Dynasty (618–907 CE). A caravan traveling along the Silk Road takes rest at an inn. At the inn, they encounter a young woman yelling at and beating an old man. A member of the caravan approached the woman and inquired why she was treating the old man in such a manner. The woman replies the man is her grandson and it's none of his business. The caravan member later learns that the woman is more than 200 years old and she was punishing her grandson for letting himself get old by not eating the flowers, known as Longevity of Life in the summer, the berries

Goji berry *(Avicmart/Wikimedia Commons)*

in the fall, and the root bark also known as the Skin and the Bone of the Earth in the winter, and the leaves also known as the Essence of Heaven in the spring, of the goji berry plant.

Another tale tells of Li Ching Yuen, an herbalist born in the Sichuan Province of China in 1677. As a young man he went up into the mountains for long periods studying and living off herbs and plants he studied on these treks. Li Ching Yuen is said to have lived 256 years. When questioned about the secret to his longevity, he answered "keep a quiet heart, sit like a tortoise, walk sprightly like a pigeon, sleep like a dog." However there are conflicting reports that he answered the secret to his longevity was eating goji berries daily.

The legend of Goji berries being a fountain of youth was further promoted by Tang Dynast poet Liu Yuxi (722–824 CE) in his poem "The Well of Youthful Living." The English translation is as follows: "A cool well beside the monk's house, a clear spring feeds the well and the water has great powers, emerald green leaves grow on the wall, the deep red berries shine like copper, the flourishing branch like a walking stick, the old root in a dog's shape signals good fortune, the Goji nourishes body and spirit, drink of the well and enjoy a long life."

In addition to longevity, goji berries have been touted as a cure for numerous medical conditions. There have been a variety of individual studies over the years on the benefits of goji berries for everything from cancer to Alzheimer's. Although these studies show potential promise, there are still an insignificant number of scientific research studies completed to prove the effectiveness of goji berries in treating those conditions and others.

Over the years, Goji berries have been touted as a cure for asthma, to prevent aging, for preventing blindness caused by diabetes, for lowering blood pressure, keeping blood sugar levels stable, and preventing cancer. It's also supposed to lower cholesterol levels, cure headaches, and unpleasant symptoms associated with menopause. Some sellers also claim it fights fatigue, makes you happy, and improves sexual function. But again, there is no evidence the goji berry is effective for treatment of these conditions. (Note that in 2007 the Australian government cracked down on two companies selling Goji juice for illegally promoting it as a treatment for sexual dysfunction, obesity, liver problems and allergies.)

PLANT HISTORY. The Goji berry plant is believed to have originated in the Ningxia Province of China. It subsequently spread to the continents of Europe, Africa, and Australia. The plant was introduced into the United Kingdom in the 1730s by the Duke of Argyll. In the United Kingdom, where the plant is often grown as a hedge, the Goji plant is therefore known as the Duke of Argyll's tea tree.

GOJI PLANT CHARACTERISTICS. A deciduous shrub, it grows long thin vines. It has thorny stems although thornless cultivars are now available. The shrub grows three to six feet tall when pruned regularly, but in its natural state can grow up to 10 feet tall. It grows approximately six feet wide when pruned regularly, or up to 13 feet in their natural state.

The shrub bears either an alternating pattern or small bundle of green ovate shaped leaves measuring up to two and three-quarter inches long and one and three-eighth inches wide. In late spring and early summer the plant bears small bell shaped purple blossoms. The shrub is self fertile, pollinated by the wind, by bees and other insects.

The plant requires full sun and grows best in alkaline soils. It's often grown as a hedge for privacy. It can also be grown in containers. Goji shrubs are propagated vegetatively and by seed. Established shrubs tolerate strong winds and are somewhat drought tolerant. They begin fruiting in their second or third year, with full fruiting by their fifth year.

▶ FAMILY: Solanaceae. GENUS: *Lycium* L. SPECIES: *Lycium barbarum* L.; *Lycium chinesis.*

SELECTING A PLANT. If you live in USDA zones 3, 4 or 10, for best success in growing a goji berry plant, select one of the newer cultivars which have been successfully grown in those zones, like the 'Phoenix Tears' cultivar. If thorns near the stem are a concern for you, be sure to select a thornless cultivar.

Goji plants that were vegetatively propagated have often been reported to start bearing fruit sooner than those grown from seed. Plants propagated from seed are not always true to type.

GROWING A GOJI BERRY PLANT. The plant grows in a variety of soil types provided there's adequate drainage. But the plant grows best in light soils like sandy loam with a pH between 6.5 and 8.2. The plant needs full sun for a good crop, but can tolerate light shade. It also grows best in nutrient rich soils.

The best time to plant a goji berry plant is in the spring before it comes out of dormancy. If needed, amend your soil with compost and products to raise its alkalinity. Dig a hole slightly deeper and larger than the pot the plant arrived in. Spread out the plants roots if needed. Place the plant in the hole and refill with soil. Water the plant thoroughly. The plant should be watered regularly to make sure the plant roots do not dry out, but be careful not to over water it during its first year. The plant should be watered regularly during fruiting. The berries are prone to blossom end rot if the plant receives uneven or insufficient moisture. The plant will benefit from a light fertilizer. More than one plant should be spaced six to eight feet apart. If you chose to plant the shrub in a planter, the planter should be a minimum of 18 inches in diameter to accommodate mature plant roots. The planter should also have adequate drainage holes.

If containing the growth of the shrub to a certain size is desired for space concerns, or to make it easier to harvest, the plant should be pruned in the winter when it is dormant. Do not prune the plant in its first year. Pruning should also be done to remove any dead wood and to allow more light to shine throughout the plant. Note that pruning will impact the crop yield for the year because new fruit is borne on the past year's wood.

The major pests of the Goji plant are mites and aphids. Birds are also frequently a problem since they enjoy eating the berries.

GROWING A GOJI PLANT FROM SEED. Since seedlings do not transplant well, each seed should be started in its own container. Fill each container with a moist potting soil and plant a seed a quarter- to half-inch deep. Cover each container with clear plastic or a clear cloche to improve germination rates. Seeds germinate best at a temperature of 65–68 degrees Fahrenheit. Place the containers indoors in an area that receives indirect sunlight. The seeds should germinate in about three weeks. Young seedlings are very sensitive to direct sunlight, so if you started them indoors, they should be slowly acclimated to direct sun.

The seedlings need to develop a strong root system before being transplanted. An indication of this is when the seedlings have several pairs of leaves.

▶ USDA HARDINESS ZONES: *Lycium barbarum* 3–10; *Lycium chinensis* 6–9. CHILLING HOURS: 0. WATER REQUIREMENTS: Average.

HOW TO CONSUME. When fully ripe, goji berries can be eaten fresh. They should be left to ripen fully on the plant. Since they bruise easily when handled ripe, the commonly recommended method of harvest is to place a sheet beneath the plant and gently shake the berries from the shrub.

Goji berries are frequently dried. They can be dried in an electric food dehydrator, in the oven or in the sun. To dry them, first wash the berries. So that they will dry faster, they should be pre-treated before drying. Pre-treating them involves dipping them into boiling water for 30 seconds to crack their skins, then placing them in ice water briefly. Drain them on paper towels then lay them out on a drying tray.

If you opt to sun dry the berries, it requires a minimum temperature above 86 degrees Fahrenheit for several days, with humidity below 60. Cover the berries on the drying tray with cheesecloth. Turn the berries every so often so they will dry evenly. Dry them until they are leathery but pliable but with no obvious moisture in the center.

After they have been sun dried, they need to be pasteurized to kill any insects or insect eggs that may have gotten on the berries. To pasteurize them, you can seal them in a freezer bag and place them in the freezer for a minimum of 48 hours at a minimum temperature of 0 degrees Fahrenheit. Or, you can spread the berries in a shallow pan and heat them in the oven for half hour at 160 degrees Fahrenheit.

If you don't want to sun dry the berries and don't have an electric food dehydrator, you can dry them in your oven. Note that it does take about twice as long to dry them in the oven than an electric dehydrator. But it takes less time than sun drying the berries.

To dry them in the oven, the oven door needs to remain open at least six inches during the entire process to provide the necessary air circulation. The temperature near the berries needs to be 140 degrees Fahrenheit. Since the oven door has to remain cracked open, you will need to place an oven thermometer near the berries and adjust the oven dial accordingly so that the thermometer reads 140 degrees Fahrenheit. The dried goji berries should be stored in a moisture proof container in a cool dark dry place. They should be used within a year.

Dried Goji berries are used in a variety of baked goods, in salads, as a topping for oatmeal, and as a snack. They are frequently used the way raisins are used. Fresh goji berries are sometimes boiled to make an herbal tea.

ADDITIONAL INFORMATION. WARNING: Medical professionals strongly advise you to consult your doctor first before eating goji berries if you take Warfarin or drugs for blood pressure or diabetes, because the berries can interact with those drugs. They also advise you to consult your doctor first before consuming goji berries if you are pregnant.

Goji berries contain polysaccharides, eight essential amino acids and are a good source of vitamins A and C, and iron, zinc, fiber and antioxidants. Because of the antioxidants in goji berries, they are also starting to be used in cosmetics. Specifically they're starting to show up in face creams meant to moisturize your skin and reduce sun damage.

REFERENCES

Amagase, H., and D.M. Nance. "A Randomized, Double-Blind, Placebo-Controlled, Clinical Study of the General Effects of a Standardized *Lycium barbarum* (Goji) Juice, GoChi." *Journal of Alternative Complementary Medicine.* Vol. 14, no. 4, May 2008, p. 403–412.

"Are Goji Berries Really the Healthiest Food on the Planet?" *Environmental Nutrition.* Vol. 33, no. 2, February 2010, p. 7.

Bucheli, P., Q. Gao, R. Redgwell, et al. Biomolecular and Clinical Aspects of Chinese Wolfberry. In: Benzie, I.F.F., and S. Wachtel-Galor, editors. *Herbal Medicine: Biomolecular and*

Clinical Aspects. 2nd edition. Boca Raton (FL): CRC Press/Taylor & Francis; 2011. Chapter 14. Available from: https://www.ncbi.nlm.nih.gov/books/NBK92756/

Bucheli, P., K. Vidal, L. Shen, Z. Gu, et al. "Goji Berry Effects on Macular Characteristics and Plasma Antioxidant Levels." *Optometry and Vision Science.* Vol. 88, no. 2, February 2011, p. 257–262.

Chaey, Christina. "All about Dried Goji Berries, the Superfood Raisin." *Bon Appetit,* June 11, 2015. www.bonappetit.com/test-kitchen/ingredients/article/goji-berries. Accessed 04/11/18.

Daugs, Don. "Business Tips for Growing Goji Berries." *Countryside & Small Stock Journal.* Vol. 99, no. 5, September/October 2015, p. 22–25.

Daugs, Don. "Grow the Alpha Superfood in Your Garden!" *Countryside & Small Stock Journal.* Vol. 98, no. 3, May/June 2014, p. 63–67.

Demchak, Kathy. "Goji Berry Culture." Pennsylvania State University, College of Agricultural Sciences Extension. https://extension.psu.edu/goji-berry-culture. Accessed 02/17/16.

Dharmananda, Subhuti. "Lycium Fruit." Institute for Traditional Medicine, Portland, OR. www.itmonline.org/arts/lycium.htm. Accessed 04/29/18.

Erickson, Kim. "Gorgeous Goji." *Better Nutrition,* vol. 71, no. 2, February 2009, p. 26.

"5 Questions About the Goji Berry Business." *Countryside & Small Stock Journal,* vol. 99, no.5, September/October 2015, p. 25.

Flannery, Dean. "Way to Goji." *Chatelaine,* vol. 79, no. 13, December 2006, p. 43.

"Go Goji Berry!" *Flex,* vol. 31, no. 6, June 2014, p. 86.

"Goji." Medline Plus, U.S. National Library of Medicine. https://medlineplus.gov/druginfo/natural/1025.html. Accessed 04/30/18.

"Goji." WebMd. www.webmd.com/vitamins/ai/ingredientmono-1025/goji. Accessed 04/11/18.

"Goji Berries." Royal Horticultural Society. www.rhs.org.uk/advice/profile?PID=581. Accessed 02/17/16.

"Goji Berry." Specialty Cropportunities. Ontario Ministry of Agriculture Food & Rural Affairs. www.omafra.gov.on.ca/CropOp/en/spec_fruit/berries/goji.html. Accessed 02/17/16.

"Goji (Lycium spp.)" Pinnacle Health, University of Pittsburgh Medical Center. www.pinnaclehealth.org/wellness-library/blog-and-staywell/health-resources/article/39646. Accessed 03/16/18.

Graham, Liz, and Lisa Guy. "Goji Berry Fights Blindness." *The Sunday Mail (Brisbane),* June 24, 2012, p. 2.

Greene, Amanda. "256 Years Old Man Breaks The Silence Before His Death And Reveals SHOCKING Secrets To The World." Parhlo.com. https://world.parhlo.com/256-years-old-man-breaks-silence-death-reveals-shocking-secrets-world/. Accessed 04/23/18.

Helmer, Jodi. "Goji Berries." *Better Nutrition.* Vol. 68, no. 9, September 2006, p. 18.

Lee, Hye Won, Young Hwa Kim, Yun Hee Kim, Gwan Ho Lee, and Mi Young Lee. "Discrimination of Lycium Chinese and Lycium Barbarum by Taste Pattern and Betaine Analysis." *International Journal of Clinical and Experimental Medicine,* vol. 7, no. 8, 2014, p. 2053–2059.

"*Lycium chinense.*" Missouri Botanical Garden. www.missouribotanicalgarden.org/PlantFinder/PlantFinderDetails.aspx?kempercode=e352. Accessed 04/29/18.

"*Lycium chinense* P. Mill." New England Wild Flower Society. https://gobotany.newenglandwild.org/species/lycium/chinense/. Accessed 04/12/18.

Marnell, Cat. "Secret Ingredient: Goji Berry." *Lucky.* Vol. 9, no. 3, March 2009, p. 137.

McLean, Tamara. "Goji Berry Could Protect Against Skin Cancer: Study." AAP Australian National News Wire, November 17, 2008.

Miles, Janelle. "Goji Health Claims Probed." *The Courier Mail (Brisbane).* Late City Edition, June 30, 2007, p. 25.

Normandeau, Sheryl. "Growing GOJI." *Gardener.* Vol. 23, no. 2, Summer 2017, p. 20.

O'Mathuna, Donal. "Can Goji Berries Help Prevent Ageing?" *Irish Times.* May 4, 2010.

Potterate, Olivier. "Goji (Lycium barbarum and L. chinense): Phytochemistry, Pharmacology and Safety in the Perspective of Traditional Uses and Recent Popularity." *Planta Medica.* Vol. 76, 2010, p. 7–19.

"Preserving Food: Drying Fruits and Vegetables." University of Georgia Cooperative Extension, College of Family and Consumer Services. 2000.

Rivera, C.A., C.L. Ferro, A.J. Bursua, and B.S. Gerber. "Probable Interaction Between Lycium Barbarum (Goji) and Warfarin." *Pharmacotherapy.* Vol. 32, no. 3, March 2012, p. 50–53.

Schalau, Jeff. "Growing Goji Berries." University of Arizona Cooperative Extension, Yavapai County. December 2, 2015. https://cals.arizona.edu/yavapai/anr/hort/byg/archive/growing goji.html. Accessed 04/27/18.

Stone, Jett. "Gaga Over Goji." *Psychology Today*, vol. 42, no. 2, March/April 2009, p. 61.

Turner, Lisa. "5 Things to Do with ... Goji Berries." *Better Nutrition*, vol.76, no. 9, September 2014, p. 14.

Weaver, Claire. "Why Goji Is More Fad Than Fact." *The Sunday Telegraph (Sydney)*, June 17, 2007, Local section, p. 19.

Weber, Christopher. "The Latest Go-To Plant." *Chicagoland Gardening*, November/December 2015, p. 21–23.

Wilson, Christian. "The Goji Juice Story." Ezine. Sept. 5, 2008. http://ezinearticles.com/?The-Goji-Juice-Story&id=1463629. Accessed 05/28/16.

Zevnik, Neil. "Goji Asia's Berried Treasure." *Better Nutrition*, vol. 69, no. 10, October 2007, p. 63–65.

Indian Gooseberry (*Phyllanthus emblica* L.)

Also known as an Amla, Emblic, and a Malacca fruit, the Indian Gooseberry is round with a smooth translucent skin. Measuring in size from an inch to about two inches, it usually has six vertical lines or furrows running from the fruit's apex to its base. Initially light green, it turns whitish-green or yellowish-green when mature. The berry is hard to the touch, and has a stone in the center with about a half dozen small seeds inside. The fruit pulp is juicy and crisp. It's very astringent when it is unripe. When ripe, its taste is sour, bitter, and acidic.

INDIAN GOOSEBERRY LORE. The Indian Gooseberry tree is considered a sacred tree in the Hindu religion. The god Vishnu is believed to dwell in the tree. According to Hindu mythology, the Indian Gooseberry tree grew from the heavenly tears shed by the god Brahma. Brahma is believed to have created the world and all its creatures. Since the Indian Gooseberry tree grew from his tears, it's believed to cure all types of ailments.

The city of Malacca in Malaysia is supposedly named after the Melaka tree (Indian Gooseberry tree). According to lore, the city was so named because the fugitive ruler Parameshwara is said to have been resting under a Melaka tree when he decided the location was perfect for establishing his new kingdom.

In Malaysia and India herbal medicine, various preparations of the Indian Gooseberry fruit, seeds, and bark are used to treat a variety of conditions. These conditions include but are not limited to indigestion, heartburn, constipation, hemorrhoids, diarrhea, arthritis, osteoporosis, and eye disorders, cardiac and urinary problems. It's also used to reduce blood sugar levels. The Indian Gooseberry is also said to improve one's hair, nails and teeth, fight cell degeneration and prevent cancer, maintain healthy liver function, and increase production of red blood cells and hemoglobin. An oil derived

Indian gooseberry *(sarangib/pixabay.com)*

from the fruit is believed to restore hair and is sometimes added to shampoos in India. However there's currently insufficient evidence regarding its effectiveness for these conditions.

TREE HISTORY. Originating in southeastern Asia, some believe specifically India; the tree is frequently found growing in India, Thailand, Singapore, and Malaysia. The Indian Gooseberry was introduced into the United States in 1901. Today the tree is also growing in the Caribbean Islands, Cuba, Puerto Rico, Panama, Central America, Philippines, Japan, Iraq and Iran.

TREE CHARACTERISTICS. The Indian Gooseberry is a small to medium sized deciduous tree measuring from 25 to 60 feet tall. It bears small alternate light green leathery oblong leaves that resemble pinnate leaves. They measure generally less than half an inch wide and less than three-quarters of an inch long. The tree's bark is smooth, flaking, and grayish-brown. The tree bears clusters of small male and female greenish yellow blossoms. The female blossoms are larger than the male ones. Occasionally the trees produce only male or female blossoms. The blossoms are pollinated by honeybees, other insects, and the wind. Cross pollination is preferred.

The trees are propagated by seed, by budding, and by air layering. Trees propagated by budding and air layering begin to fruit after five or six years, whereas trees grown from seed may take up to 10 years before beginning to bear fruit. On the India Biodiversity Portal the tree's life expectancy is listed as 15 years, however, Julia Morton's *Fruits of Warm Climates* lists the tree as producing for 50 years. The trees bear more than 30 pounds of fruit annually. Cultivars include but are not limited to the 'Banarsi' which produces fruit early in the season, the 'Chakaiya' which is a prolific fruit producer, and the 'Francis,' which is a consistent fruit producer.

► FAMILY: Euphorbiaceae. GENUS: *Phyllanthus* L. SPECIES: *Phyllanthus emblica* L.

SELECTING A TREE. When selecting a tree from the nursery, check the tree's overall health. Look for active growth, such as new or young leaves. Make sure its leaves are not yellow or discolored, and free from insects or signs of insect damage. Check the entire tree to make sure it is free from injury, such as broken branches, cuts in the trunk, etc. And if possible, gently lift the tree out of the pot to check the roots. Overgrown trees that have seriously girdled roots are likely to develop a poor root structure so they should be avoided.

HOW TO GROW AN INDIAN GOOSEBERRY TREE. The Indian Gooseberry tree prefers full sun and slightly acidic to slightly alkaline soils. It prefers a pH from six to eight, but will tolerate from a five to eight. The tree requires well draining deep soil rich in organic matter. It grows in soils ranging from sandy loam to light clay. If you are planting more than one tree, they should be spaced 30 feet or more apart.

Once you've selected the best location for the tree, dig a hole at least twice the diameter of the pot the tree came in, about as deep as the container. Place the tree in the hole, loosening or cutting away any girdled roots. The tree should be planted so that the top of the root ball is about one to two inches higher than the soil level. Refill the hole with the original soil. Water thoroughly. The tree should be watered regularly until it becomes established. Established mature trees only need to be watered two to three times during the summer time.

HOW TO GROW FROM SEED. To grow a tree from seed, remove the stone from a gooseberry and crack it open to reveal the seeds inside. Next perform a float test. Place the seeds in a container of water overnight. Those seeds that sink to the bottom are said to have nearly 100 percent germination rate. Remove the seeds that sunk to the bottom and plant each one about a quarter-inch deep in a pot filled with an equal mix of potting soil and compost. Keep the soil moist. The seeds germinate best in temperatures around 80 degrees Fahrenheit. They typically germinate in two to seven weeks.

▶ USDA HARDINESS ZONES: 10–11. CHILLING HOURS: 0. WATER REQUIRE-MENTS: Prefers 60–98 inches rainfall annually, but tolerates 28–165 inches.

HOW TO CONSUME. Indian Gooseberries are generally not eaten fresh as a fruit due to its acidic bitter taste. The fresh fruit is sometimes added to dishes which need acidity. It's frequently cooked with sugar and saffron. It can is used in baked goods such as tarts. If can also be candied and used in preserves. The Gooseberries are often dried and powdered for use in herbal remedies.

ADDITIONAL INFORMATION. The fruit is rich in antioxidants. The tree's leaves are used as fodder for cattle. Because they're believed to correct excessive alkaline soils, they are also sometimes used as green manure. The tree wood is believed to clarify water so it is sometimes used in wells. And wood chips are sometimes thrown in muddy streams to clarify the water. In Malaysia a fruit extract has also been used as a hair dye. In some regions of the world the tree is considered an invasive species.

REFERENCES

Aggarwall, Vandana. "Malacca, The Historical City of Malaysia." *Woman's Era*, Sept. 2, 2017, p. 30–31.
"Amla, the Ancient Healing Fruit." *The Statesman.* August 17, 2017. www.thestatesman.com/lifestyle/amla-ancient-healing-fruit-1502439133.html. Accessed 12/18/18.
Cunningham, Sally. *Asian Vegetables: A Guide to Growing Fruit, Vegetables and Spices from the Indian Subcontinent.* Bath (UK): Eco-Logic Books. 2009.
Fern, Ken. "*Phyllanthus emblica* L." Tropical Plants Database. http://tropical.theferns.info/viewtropical.php?id=Phyllanthus+emblica. Accessed 12/13/18.
Ganesan, R. and R. Siddappa Setty. "Regeneration of *Amla*, an Important Non-timber Product from Southern India." *Conservation & Society*, vol. 2, no. 2, 2004, p. 365–375. http://dlc.dlib.indiana.edu/dlc/bitstream/handle/10535/3054/regeneration_of_amla.pdf?sequence=1&isAllowed=y. Accessed 12/13/18.
"Indian Gooseberry." WebMd. www.webmd.com/vitamins/ai/ingredientmono-784/indian-gooseberry. Accessed 12/19/18.
Khoo, Hock Eng, et al. "Phytochemicals and Medicinal Properties of Indigenous Tropical Fruits with Potential for Commercial Development." *Evidence-Based Complementary and Alternative Medicine*, vol. 2016, p. 1–20. www.ncbi.nlm.nih.gov/pmc/articles/PMC4906201/. Accessed 11/02/18.
"Learn How to Grow Indian Gooseberry." Balcony Garden Web. https://balconygardenweb.com/growing-amla-tree-how-to-grow-indian-gooseberry/. Accessed 12/13/18.
Leon, Anastasia. "How to Grow Indian Gooseberry from Seed." SFGate. http://homeguides.sfgate.com/grow-indian-gooseberry-seed-76321.html. Accessed 03/22/16.
Lynch, S. John, and Fred J. Fuchs, Sr., "A Note of the Propagation of Phyllanthus Emblica L." *Proceedings of the Florida State Horticultural Society*, vol. 68, 1955, p. 301–302. https://fshs.org/proceedings-o/1955-vol-68/301-302%20(LYNCH).pdf. Accessed 12/13/18.
Morton, Julia F. *Fruits of Warm Climates.* Eugene, OR: Wipf and Stock Publishers, 2003.

Orwa, C., et al. "Emblica officinalis." 2009 Agroforestree Database: A Tree Reference and Selection Guide Version 4.0. www.worldagroforestry.org/treedb/AFTPDFS/Emblica_officinalis.PDF. Accessed 12/19/18.

Pandey, Mahendra. "An Amla a Day." *Down to Earth*, June 11, 2015. www.downtoearth.org.in/coverage/an-amla-a-day-1208. Accessed 12/19/18.

"*Phyllanthus emblica* L." Broome, R., Sabir, K., Carrington, S. "Plants of the Eastern Caribbean." 2007. University of the West Indies, Cave Hill Campus, Barbados. http://ecflora.cavehill.uwi.edu/plantdetails.php?pid=930&sn=Phyllanthus+emblica&cn=Indian+gooseberry&gh=tree+or+tree-like. Accessed 12/13/18.

"*Phyllanthus emblica* L." India Biodiversity Portal. https://indiabiodiversity.org/species/show/31625. Accessed 12/13/18.

"*Phyllanthus emblica* L." Plants for a Future. https://pfaf.org/user/Plant.aspx?LatinName=Phyllanthus+emblica. Accessed 12/10/18.

"*Phyllanthus emblica* L." USDA, NRCS. 2018. The PLANTS Database. National Plant Data Team, Greensboro, NC 27401-4901. https://plants.usda.gov/core/profile?symbol=PHEM2. Accessed 12/20/18.

"*Phyllanthus emblica* L." World Wide Fruits. www.worldwidefruits.com/phyllanthus-emblica-indian-gooseberry.html. 12/19/18.

Pushpakumara, D., et al., editors. *Underutilized Fruit Trees in Sri Lanka.* New Delhi: World Agroforestry Center, 2007.

"Synopsis of Emblica." University of California, Davis. https://herbarium.ucdavis.edu/manuscripts/webster/New%20World/2002%20Synopsis%20of%20Phyllanthus%20subgenus%20Emblica.pdf. Accessed 12/13/18.

Thankitsunthorn, S., et al. "Effects of Drying Temperature on Quality of Dried Indian Gooseberry Powder." *International Food Research Journal*, vol. 16, 2009, p. 355–361. www.ifrj.upm.edu.my/16%20(3)%202009/8[1]%20George.pdf. Accessed 12/14/18.

Jackfruit (*Artocarpus heterophyllus* Lam.)

The jackfruit, also known as a jak, is the largest cultivated tree-borne fruit. An oblong shaped fruit, it ranges from 16 to 36 inches long. It's often compared to the size of a large watermelon. A jackfruit can be very heavy. Although it can weigh up to 80 pounds or more, most of the fresh jackfruit you'll find in local specialty markets generally weigh 10 to 20 pounds. (Let's face it, how many of us are able and willing to carry a 50 plus pound fruit home?)

The outer skin or rind of the fruit is hard, green or pale yellow when ripe, and covered with little cone-like points. The fruit has a white pith and yellow pulp in bulb like segments. Each bulb of flesh contains a seed of up to one and one-half inches long. The raw seeds are inedible, but can become edible if cooked.

A few days before the jackfruit is fully ripe, it emits a slight aroma, or what some consider an odor, which has been compared to that of an armpit. When ripe, the flesh is said to taste something like a blend of a pineapple and a banana. Some people compare the taste of a ripe jackfruit to that of Juicy Fruit gum. But when it's not ripe, its texture and taste is often compared to pulled pork, which is why jackfruit is often marketed as a vegetarian substitute for meat. Many consumers describe the unripe taste of jackfruit as simply a

Jackfruit *(vasanthkumar/pixabay.com)*

starchy food which can take on the taste of whatever taste the cook decides to flavor it as.

JACKFRUIT LORE. In Thai folklore, the seeds of the jackfruit are valued and often strung together as a necklace or bracelet. They are believed to have the power to protect the wearer from harm, specifically from sharp objects, and from firearms exploding near the wearer.

Jackfruit is also used for folk medicine purposes in some Asian cultures. Consumption of the pith supposedly induces an abortion. The pulp is used as a cooling tonic and to treat diabetes.

In Malaysia, the sap of the jackfruit tree is used to treat snake bites. The tree root is used to treat diarrhea, fever, skin conditions and intestinal worms. And a solution of jackfruit leaf ash and coconut oil is applied to wounds and ulcers. The roasted jackfruit seeds are often regarded in some Asian countries as an aphrodisiac.

In the Thoubal district of Manipur, India is the Sacred Jackfruit Tree. The Sacred Jackfruit Tree is a historical site sitting on a small hill in Kaina. According to legend, in the 18th century the Hindu god Krishna appeared to Jai Singh Maharaja, also known as Rajarshi Bhagya Chandra (who ruled

Cut jackfruit *(pixabay.com)*

Manipur at the time) in a dream. Krishna told him to carve images of him in the Jackfruit tree. The site is where Jai Singh Maharaja had seven images of Krishna from the Jackfruit tree. The seven images were then distributed to temples in Manipur and Assam.

TREE HISTORY. The jackfruit tree dates back to ancient times. Roman author and philosopher Pliny the Elder mentioned the jackfruit in his writings. Experts believe the tree originated in the rainforests of the Western Ghats, a range of hills that extends along the west coast of India from the Tapti River in the North to the Southern tip of India. It subsequently spread to Southeast Asia and China. It was introduced into Jamaica in the late 1700s and into Hawaii and Florida in the 19th century. It was introduced in Brazil in the mid–19th century. Jackfruits are also grown in Africa and Australia.

TREE CHARACTERISTICS. The jackfruit tree is a fast growing tropical evergreen tree, and is both frost and drought sensitive. Growing up to five feet per year, the tree can reach up to 65 feet or taller, with a trunk up to two and a half feet in diameter. Mature trees bear a spreading domed canopy. The tree possesses a sticky white latex. The tree's leaves are somewhat oval shaped, glossy, dark green and leathery. They measure approximately four to six inches long.

Jackfruit trees bear both male and female blossoms. Male blossoms appear in oblong clusters, while female blossoms are borne in round clusters. Mature trees can produce up to 150 fruits per year. The tree is predominantly propagated by seed. However, it is also vegetatively propagated using root and stem cuttings. In Southeast Asia trees are also commonly propagated by grafting and budding. Trees propagated by seed can take up to eight years to begin bearing fruit. Vegetatively propagated trees begin bearing fruit after three years.

▶ FAMILY: Moraceae. GENUS: *Artocarpus* J.R. Forst. & G. Forst. SPECIES: *Artocarpus heterophyllus* Lam.

SELECTING A TREE. Since there are many jackfruit cultivars to choose from, the taste of your favorite cultivar should be a factor in your selection. Seedlings grown from seed tend to be hardier than those that have been grafted. But remember that trees grown from seed are always not true to type. The first generation of seedlings only retains about 90 percent of the characteristics of the parent tree. Grafted trees tend to be truer to the cultivar.

When selecting a tree from the nursery, check the tree's overall health. Look for active growth, such as new or young leaves. Make sure its leaves are not yellow or discolored, and the tree is free from insects or signs of insect damage. Check the entire tree to make sure it is free from injury, such as broken branches, cuts in the trunk, etc. And if possible, gently lift the tree out of the pot to check the roots. If the roots are overgrown in the pot and are seriously girdled, the tree should be avoided since it will likely develop a poor root structure.

GROWING A JACKFRUIT TREE. Jackfruit trees need full sun, so be sure to select a location with full sun that can accommodate a fully mature tree. The tree needs rich, deep, well draining soil, like clay loam or sandy soils. The soil should be moist, but not wet since the tree does not withstand wet roots. It prefers a pH of 6.0 to 7.5.

To plant a jackfruit tree, dig a hole twice the diameter of the root ball, and at least as deep as the root ball. Carefully place the tree in the hole, making sure the root graft line (if a grafted tree) is above the soil. Refill the hole with the original soil. Water thoroughly. The tree benefits from fertilization starting in its first year. But there appears to be no consistent recommendation on the frequency (ranging from every eight weeks to every six months) and type of fertilizer.

Pruning should be done to remove any dead wood. In two year old trees branches are often pruned to the first lateral branch to encourage a spreading canopy and to slow upward growth to make it easier to harvest. Potential diseases include root rot (caused by *Phytophora* or *Fusarium*). Other potential problems include blossom and fruit rot.

Pests include scales, aphids and mealy bugs. Other pests include wood boring insects. Bark borers and shoot borers can attack weak or dead tree wood.

HOW TO GROW FROM SEED. To grow from seed from a jackfruit, clean and wash away any fruit pulp from the seed. If you don't plan on planting the seed immediately, note that seeds loose their viability after a month. But if you store the seed in a damp airtight plastic container at 68 degrees Fahrenheit the seed viability could be extended for up to three months.

When you are ready to plant the seed, soak it in water for 24 hours. Removing the seed casing/shell will improve the seed germination. And since the seedling roots do not like being disturbed, selected a pot large enough so frequent transplanting into larger pots is minimized. Place the seed hilum side down, or if you're unsure which is the hilum side, just place the seed flat, in rich well draining soil, keeping the soil moist. Note the seed needs a minimum temperature of 80 degrees Fahrenheit to germinate. The seed should germinate in three to eight weeks. Grow the seedling under shade. The seedlings have a delicate taproot so they should be transplanted within their first year.

▶ USDA HARDINESS ZONES: 9–11. CHILLING HOURS: 0. WATER REQUIREMENTS: 59 inches rainfall annually.

HOW TO CONSUME. You can eat ripe jackfruits fresh. (Hint: The smell is one indication it's ripe.) You may want to wear gloves to protect your hands when handling the fruit. To eat it fresh, place the fruit on a cutting board. Coat the blade of your knife with vegetable oil to prevent the fruit from sticking to the blade. Cut through the fruit lengthwise and slice the fruit in half. Then slice the halves into two so that you have four quarters. Cut and remove the center core. Pull apart each fleshy bulb of fruit. Remove the seeds and discard any fibrous materials surrounding the seed. Fresh ripe jackfruit pulp can also be added to fruit smoothies and salads.

You can freeze any of the fresh flesh for consumption later. Simply place it in a freezer bag or container and store in your freezer at zero degrees Fahrenheit or colder. Make sure you remove the seeds and any surrounding fibrous materials from the flesh before freezing. The fruit can also be dried, made into chutney or jam. Commercially the fruit is also sold canned in brine, and syrup, and occasionally sold pickled.

Raw jackfruit seeds cannot be eaten because they contain a powerful trypsin inhibitor, which makes the seeds indigestible. But it's destroyed by heat. So jackfruit seeds need to be boiled or baked before consumed. The taste of a roasted or baked jackfruit seed is often compared to a bland chestnut.

To boil the seeds, boil them in a pot for 30 minutes or until they are soft enough to be pierced by a fork. Remove the pot from heat and drain. Remove and discard the seed shell and enjoy the boiled jackfruit seeds. To roast the seeds, lay them out in a single layer on a baking pan. Roast for a minimum of 20 minutes at 400 degrees. Let cool, then remove the outer seed shell and discard.

Unripe jackfruit has a stringy texture which is why it is often described as similar to pulled pork. In terms of cooking unripe jackfruit, cooking it is very similar to cooking a starchy food. One common method of using jackfruit is to make a barbecue jackfruit sandwich. One recipe for this can be found on the Michigan State University website at http://health4u.msu.edu/recipes/barbecue-jackfruit-sandwich. You can find numerous other jackfruit recipes online.

ADDITIONAL INFORMATION. Jackfruit tree wood is termite resistant and used in home building, and used to build furniture. The fruit skin and leaves are used as cattle feed. In March of 2018 the jackfruit was declared the official fruit of the state of Kerala, in South India. About 500 million jackfruits grow in Kerala. The state also produces jackfruit chips which are highly popular in the region and beyond. India is the largest producer of jackfruit, producing 1.4 million tons, followed by Bangladesh at 926 tons, followed by Thailand at 392 tons and Indonesia at 340 tons.

REFERENCES

Aman, Rukayah. *A Guide to Jackfruit Cultivation.* Mardi, Serdang: The Archives of the Rare Fruit Council of Australia. 1984.

"*Artocarpus heterophyllus* Lam." USDA, NRCS. 2018. The PLANTS Database. National Plant Data Team, Greensboro, NC 27401-4901. https://plants.sc.egov.usda.gov/core/profile?symbol=ARHE2. Accessed 12/26/18.

Baum, Isadora. "What Is Jackfruit? Here's Everything You Need to Know About the Trendy Meat Substitute." *Men's Health,* February 6, 2018. www.menshealth.com/nutrition/a19547142/what-is-jackfruit-how-to-eat-it/. Accessed 05/28/18.

Benwick, Bonnie S. "Read This Before You Lug Home That Jackfruit." *Washington Post,* June 6, 2016. www.washingtonpost.com/lifestyle/food/read-this-before-you-lug-home-that-jackfruit/2016/06/06/7d45b17e-282f-11e6-b989-4e5479715b54_story.html?noredirect=on&utm_term=.f6d98ad98b0b. Accessed 06/05/18.

Crane, Jonathan H., Carlos F. Balerdi, and Ian Maguire. *Jackfruit Growing in the Florida Home Landscape.* HS882. University of Florida, Institute of Food and Agricultural Sciences Extension. 2016. https://edis.ifas.ufl.edu/pdffiles/MG/MG37000.pdf. Accessed 07/01/18.

"Cultural Practices for Jackfruit." Cropsreview.com. www.cropsreview.com/growing-jackfruit.html. Accessed 11/04/15.

Cunningham, Sally. *Asian Vegetables: A Guide to Growing Fruit, Vegetables and Spices from the Indian Subcontinent.* Bath, UK: Eco-logic Books, 2009.

Fairchild, David. *The Jack Fruit (Artocarpus integra, Merrill): Its Planting in Coconut Grove, Florida.* Florida Plant Immigrants. Occasional Paper No. 16. Coconut Grove, FL: Fairchild Tropical Garden, 1946.

Freier, Lizzy. "Jackfruit Takes the Spotlight." *Restaurant Business*, vol. 116, no. 2, February 2017, p. 48.

"How to Grow Jackfruit." https://plantinstructions.com/tropical-fruit/how-to-grow-jackfruit/. Accessed 05/28/18.

"Jackfruit, a Promising Cash Crop." Cropsreview.com www.cropsreview.com/jackfruit.html. Accessed 05/26/18.

"Jackfruit Declared Kerala's Official Fruit." *South Asian Post*. March 29, 2018, p. 17.

Love, Ken, and Robert E. Paull. "Jackfruit." F_N-19. Manoa, HI: College of Tropical Agriculture and Human Resources, University of Hawaii at Manoa. 2011. www.ctahr.hawaii.edu/oc/freepubs/pdf/F_N-19.pdf. Accessed 05/28/18.

McKay, Gretchen. "Jackfruit Is a Fibrous Meat Alternative." *Pittsburgh Post-Gazette*, July 26, 2017.

Morton, Julia F. *Fruits of Warm Climates*. Eugene, OR: Wipf and Stock Publishers, 2003.

Piper, Jacqueline M. *Fruits of South-East Asia: Facts and Fiction*. Oxford, NY: Oxford University Press, 1989.

Popenoe, Wilson. *Manual of Tropical and Subtropical Fruits: Excluding the Banana, Coconut, Pineapple, Citrus Fruits, Olive and Fig*. NY: Macmillan Co., 1920.

Prakash, Om, Rajesh Kumar, Anurag Mishra, and Rajiv Kupta. "Artocarpus heterophyllus (Jackfruit): An Overview." *Pharmocognosy Reviews*, vol. 3, no. 6, July 2009, p. 353–358.

"Sacred Jackfruit Tree, Kaina Hill, Thoubal, Manipur." www.apnisanskriti.com/temple/sacred-jackfruit-tree-kaina-hill-thoubal-manipur-3981. Accessed 06/05/18.

Silver, Marc. "Here's the Scoop on Jackfruit, a Ginormous Fruit to Feed the World." National Public Radio. May 1, 2014. www.npr.org/sections/thesalt/2014/05/01/308708000/heres-the-scoop-on-jackfruit-a-ginormous-fruit-to-feed-the-world. Accessed 05/28/18.

Silver, Marc. "Whatever Happened To … The Plan to Jazz Up Jackfruit?" National Public Radio. September 3, 2017. www.npr.org/sections/goatsandsoda/2017/09/03/547337695/whatever-happened-to-the-plan-to-make-jackfruit-more-popular. Accessed 05/28/18.

Stukin, Stacie. "Serving Up Jackfruit." *National Geographic*, vol. 230, no. 6, December 2016, p. 14.

"What to Know About Jackfruit, the Next Big Thing in Produce." The Food Network. May 2016. www.foodnetwork.com/fn-dish/news/2016/05/what-to-know-about-jackfruit-the-next-big-thing-in-produce. Accessed 05/28/18.

Worley, Sam. "Everything You Need to Know About Jackfruit, the Latest Miracle Food." *Epicurious*, June 24, 2016. www.epicurious.com/ingredients/facts-tips-recipe-ideas-jackfruit-vegan-miracle-food-article. Accessed 05/28/18.

Japanese Raisin (*Hovenia dulcis* Thunb.)

Also known as a Chinese raisin and Oriental raisin, the Japanese raisin is technically not the fruit produced by the Raisin tree. The Japanese raisin is actually the fruit stalk, which turns reddish brown, swells up and has a knobby look when the "fruit" is ripe. The edible stalk, or raisin, is only about the size of an actual raisin. They're said to taste like an apple, pear or raisin. But when allowed to fully "ripen" on the tree until they fall off naturally, their texture and taste is described as being like a crunchy raisin. Although it's only the size of a raisin, the tree produces abundant amounts of them. The actual fruit of the tree is small (less than half-inch in size), hard, and initially green but it turns brown when mature. It's dry and contains three seeds. The actual fruit is not edible.

JAPANESE RAISIN LORE. In Asia, *Hovenia dulcis* has been used to treat drunkenness, hangovers, and liver damage from alcohol consumption. Studies have been done in rats which show that rats that were given a hot water extract of *Hovenia dulcis* showed a faster alcohol reduction in blood alcohol than in the control group. And a study on mice showed it helped them metabolize alcohol faster. Two studies in rats have also shown that *Hovenia dulcis* reduced and protected the rats from alcohol induced liver damage. It's the dihydromyricetin, a compound found in Japanese raisin trees that's credited for these results. But as noted, these studies have been conducted on rats, not humans. There's currently a lack of significant studies of the effects of *Hovenia dulcis* on humans.

In Korea, beverages containing dihydromyricetin are sold as hangover cures. But here in the United States the Food and Drug Administration has not approved any beverages containing dihydromyricetin to be sold as hangover cures. In Asian herbal medicine, the bark of Japanese raisin trees has been used to treat cancer. The Japanese raisins are also used to "quiet the stomach," and as a laxative.

TREE HISTORY. The tree originated in China in pre–Confucian times. It spread to Japan, Korea, Thailand, Vietnam and India. It was introduced into the West around 1820 and into Brazil around 1987. It currently is also growing in Russia, Australia, New Zealand, and the United States. In the South American rain forests and in the Tanzania area of Africa the tree has become invasive.

TREE CHARACTERISTICS. Most cultivated Japanese raisin trees grow to about 15 to 25 feet high, with a somewhat spreading round to oval crown that can be as wide as 12 feet. The tree has an erect stem and gray bark on young trees which can be peeled away to expose a dark brown fissured bark on older trees. A deciduous tree, it features large oval shaped leaves up to six inches long and up to five inches wide. The leaves are somewhat glossy and green on the top side, and pale green and slightly hairy on the underside.

The tree bears clusters of small creamy white or greenish purple blossoms. The blossom clusters are often described as "showy." The blossoms have a pleasant fragrance. The tree is self-fruitful.

The tree is more often planted for its landscape value than for its edible value. It's a low maintenance tree that basically prunes itself. The lower branches drop off as the tree grows taller. The tree has no known major pests or diseases. And it's fairly drought tolerant.

The tree is propagated by seed and vegetatively (by root and softwood cuttings). And there are no known cultivars of the tree. Trees grown from seed usually begin bearing "raisins" in seven to ten years.

Japanese raisin *(Bluesnap/pixabay.com)*

► Family: Rhamnaceae. Genus: *Hovenia* Thunb. Species: *Hovenia dulcis* Thunb.

Selecting a Tree. Starting out with a healthy tree is the best way to help improve your chances of successfully growing a Japanese raisin tree. When selecting a tree from the nursery, check the tree's overall health. Make sure its leaves are not discolored, and are free from insects or signs of insect damage. Check the entire tree to make sure it is free from injury, such as broken branches, cuts in the trunk, etc. Check to make sure the tree trunk is not sunburned. Also visually inspect the pot soil to make sure there is no fungi growing from the soil, indicating uneven or over watering that can affect the overall health of the trees. If possible, check the tree roots. Gently lift the tree out of the pot. The roots should flare out evenly from the trunk. If the roots are girdled, or kinked around in a circle, the tree is overgrown and may likely develop a poor root structure when planted in the ground.

How to Grow a Raisin Tree. The tree tolerates many types of soils, including, clay, sandy and loam soils, but prefers well draining soil. It thrives in sandy loam soil. It requires a soil pH of 6 to 7.8. The tree grows in full sun or partial shade, but fruits better in full sun. The tree's roots also need room for expansion as it grows. So take these two items into consideration when selecting a planting site. To plant a tree, dig a hole at least twice the diameter of the pot the tree came in, about as deep as the container. Place the tree in the hole, loosening or cutting away any girdled roots. The tree should be planted so that the top of the root ball is about one to two inches higher than the soil level. Refill the hole with the original soil. Water the newly planted tree thoroughly.

How to Grow from Seed. Japanese raisin seeds have an impermeable cover that makes germination difficult. So seeds should be scarified, by either rubbing the seed between sandpaper or nicking it with a file or knife, and then soaked in hot water (approximately 140 degrees Fahrenheit) for several days.

Note that it's been reported that in laboratory tests, germination was increased by 40 percent when seeds were scarified in concentrated sulfuric acid for an hour prior to sowing, or scarifying in acid plus three months of cold stratification. But sulfuric acid is not something people have around their homes. Once the seed has been scarified, cover it lightly with soil, keeping it moist, and place it in a sunny location. The seed germinates best when the soil temperature is between 59 to 68 degrees Fahrenheit. The germination can be very slow, taking anywhere from a week to several months.

▶ USDA Hardiness Zones: 6–10. Chilling Hours: 0. Water Requirements: Average.

How to Consume. The Japanese raisin can be eaten fresh or it can be cooked. If you're picking them off the tree, you can tell they are "ripe" when they swell and turn a translucent reddish brown. As they ripen, the sugars increase and it develops an apple pear-like or raisin-like flavor. But unlike raisins, they have a somewhat crunchy texture. When they are fully ripe, they fall off the tree. And in Asia they frequently wait until they fall off the tree to harvest them. Be sure to thoroughly wash them before consuming.

Fresh Japanese raisins can be added to salads, cereal, and other dishes that you would normally add dried fruits or raisins. They can also be heat dried. Dried Japanese raisins reportedly can be stored for a few months in a dark dry place. You can use Japanese raisins in breads and other baked goods, including cookies and pies.

Additional Information. The tree wood is sturdy and heavy and used to make furniture. In Asia an extract of the Japanese raisin trees seeds, leaves and branches is used as a sweetener, a substitute for honey and is called "tree honey."

REFERENCES

"Feeling Foggy? Japanese Raisin Is Good for What 'Ales' You." [Broadcast Transcript] National Public Radio. May 25, 2016. www.npr.org/2016/05/25/479420372/feeling-foggy-japanese-raisin-is-good-for-what-ales-you. Accessed 06/15/18.

"Feeling Fruity?" *Manawatu Standard*, January 27, 2012, p. 21.

Gilman, Edward F., and Dennis G. Watson. "*Hovenia dulcis* Japanese Raisintree." ENH455. University of Florida, Institute for Food and Agricultural Sciences. 2014. http://edis.ifas.ufl.edu/pdffiles/ST/ST29600.pdf. Accessed 11/27/18.

Great Soviet Encyclopedia, 3rd edition. Gale Group. 1979.

"Hovenia, Also Called Chinese, Japanese or Oriental Raisin Tree." The Rare Fruit Club, Western Australia. www.rarefruitclub.org.au/Level2/Hovenia.htm. Accessed 06/17/18.

"Hovenia dulcis." North Carolina State University, Cooperative Extension. https://plants.ces.ncsu.edu/plants/all/hovenia-dulcis/. Accessed 06/14/18.

"Hovenia dulcis." Plants for a Future. www.pfaf.org/user/plant.aspx?LatinName=Hovenia+dulcis. Accessed 06/15/18.

"*Hovenia dulcis*." USDA, NRCS. 2018. The PLANTS Database. National Plant Data Team, Greensboro, NC 27401-4901. https://plants.usda.gov/core/profile?symbol=HODU2. Accessed 06/12/18.

"Hovenia dulcis (Japanese Raisin Tree)." BioNET-International. https://keys.lucidcentral.org/keys/v3/eafrinet/weeds/key/weeds/Media/Html/Hovenia_dulcis_(Japanese_Raisin_Tree).htm. Accessed 06/13/18.

"*Hovenia dulcis*—Japanese Raisin Tree." University of Alabama in Huntsville. https://www.uah.edu/facilities-and-operations/facilities/grounds/trees/deciduous/157-facilities-operations/2393-grounds-trees-japanese-raisin-tree. Accessed 06/15/18.

Hyun, Tae Kyung, Seung Hee Eom, Chang Yeon Yu, and Thomas Roitsch. "*Hovenia dulcis*—An Asian Traditional Herb." *Planta Medica*, vol. 76, 2010, p. 943–949.

"Japanese Raisin Tree." Incredible Edibles (Tharfield Nursery Ltd). www.edible.co.nz/fruits.php?fruitid=36_Japanese%20Raisin%20Tree. Accessed 06/17/18.

"Japanese Raisin Tree." Specialty Produce. www.specialtyproduce.com/produce/Japanese_Raisin_Tree_9497.php. Accessed 06/15/18.

"Japanese Raisin Tree—Tasty Twigs!" Just Fruits and Exotics. www.justfruitsandexotics.com/JapaneseRaisinTreeArticle.pdf Accessed 06/15/18.

Kitsteiner, John. "Permaculture Plants: Raisin Tree." Temperate Climate Permaculture. August 12, 2014. http://tcpermaculture.com/site/2014/08/12/permaculture-plants-raisin-tree/. Accessed 6/13/18.

Koller, Gary, and John H. Alexander III. "The Raisin Tree—Its Use, Hardiness and Size." *Arnoldia* (Magazine of the Arnold Arboretum at Harvard University), vol. 39, no. 1, 1979, p. 7–15.

Michaels, Francis. "Japanese Raisin Growing Information." Green Harvest. http://greenharvest.com.au/SeedOrganic/FruitTrees/JapRaisinGrowingInformation.html. Accessed 06/14/18.

O'Keefe Osborn, Corinne. "What Is Hovenia Dulcis?" Healthline. October 26, 2017. https://www.healthline.com/health/hovenia-dulcis. Accessed 05/21/18.

Ortho All About Citrus and Subtropical Fruits. Des Moines, IA: Meredith Books, 2008.

Park, J.S. "Evaluation of Herb-Drug Interactions of Hovenia dulcis Fruit Extracts." *Pharmacognosy Magazine*, vol. 13, no. 50, April—June 2017, p. 236–239.

Parmar, Chiranjit, Editor. "Japanese Raisin." *Fruitipedia*. www.fruitipedia.com/japanese_raisin.htm. Accessed 06/15/18.

Plant Files: Japanese Raisin Tree, Honey Tree." Dave's Garden. https://davesgarden.com/guides/pf/go/21/. Accessed 06/15/18.

"*Raisin Tree Fact Sheet*." California Rare Fruit Growers. 1996. www.crfg.org/pubs/ff/raisintree.html. Accessed 06/15/18.

Reich, Lee. "If You Like the Taste of Raisins, Why Not Grow a Raisin Tree?" *The Canadian Press*, March 24, 2003.

Simoons, Frederick J. *Food in China: A Cultural and Historical Inquiry*. Boca Raton, FL: CRC Press, 2014.

Sternlicht, Alexandra. "This 27-Year Old Ex-Tesla Manager Wants to Save You from Your Hangover." *Forbes.* June 14, 2018. www.forbes.com/sites/alexandrasternlicht/2018/06/14/this-27-year-old-ex-tesla-manager-wants-to-save-you-from-your-hangover/#3f00c 97838e3. Accessed 06/15/18.

Wrigglesworth, Jane. "Add Edible Variety—And Enjoy." *Waikato Times*, Feb. 4, 2012, p. YW18.

Java Apple (*Syzygium samarangense* [Blume] Merr. & L.M. Perry)

Also known as a wax apple, wax jambu, and a rose apple, the Java apple is a bell shaped smooth waxy skinned fruit about one and a half to three inches long. Its outer skin is thin and ranges in a variety of colors, including white, light green, pink, red, deep purplish red, to almost black. The fruit pulp is white, low acid and mildly sweet with a slight rose aroma. The pulp is generally juicy, with pink skinned fruit usually being the juiciest. Despite its name, it tastes more like a pear than an apple. The fruit's texture is frequently compared to being similar to that of a watermelon and similarly juicy. The pulp is dense on the outside but fluffy and spongy at its core. The fruit can have no seeds, or up to three round seeds measuring one centimeter in diameter at the core of the fruit.

The Java apple is similar in appearance to the Malay apple, but they are not the same. Although both are in the same botanic genus and family, they are different species.

JAVA APPLE LORE. There's a popular tale in the Philippines regarding how the Java apple (known there as a Macopa tree) came into existence. According to the tale there was a small town which in some versions was located near the mountains, and in others located by the coast. The town possessed a golden bell which supposedly appeared out of nowhere. But since the appearance of the bell, the town enjoyed plentiful harvests and happiness, which they attributed to the bell. When the Friars arrived and the first church was built, the townspeople decided the bell should be hung in the church. The bell had a unique pleasant tone. When rung, it would remind the people of how they were blessed with a good life. News of the miraculous bell spread to other areas, and unfortunately nearby bandits heard of the gold bell.

Somehow the townspeople got word that bandits were on their way to steal the bell. So the Friar and a handful of townspeople decided to hide the bell. So they buried it. When the bandits arrived, they were outraged the bell was gone. They tortured the priest and townspeople they found in the church to compel them to tell them where the bell was. But they refused, so the bandits killed them. Upset they could not locate the bell, the bandits left.

Since the only people who knew where the bell was hidden had been

Java apple *(gkgegk/pixabay.com)*

murdered, the townspeople could not retrieve the bell. As a result their harvests began to dwindle. The townspeople fell into despair and resigned themselves to a life of misery.

Years later a strange tree grew beside the church that bore strange bell shaped fruit. Then one night a villager had a compelling dream summoning him to dig near the tree with the bell shaped fruit. When he told the townspeople of this dream, they grabbed their shovels and began to dig near the Java apple tree. As soon as they did, they hit something metal. It was the golden bell. The tree with the bell shaped fruit had grown from the spot where the bell was hidden as a sign to the townspeople. The bell was dug out of the ground and once again hung in the church where the bell tolled again. The harvests once again became plentiful and people were happy again. Soon the townspeople discovered the bell shaped fruit on the tree was edible. The fruit looked like a cup so they named it a "macopa" which translates into "cup" in the Philippines. And that is how the Java apple tree came into existence.

TREE HISTORY. The Java apple originated in the area from Malaysia to the Adaman and Nicobar Islands in India. Portuguese traders carried it from India to East Africa. It spread to the Philippines, Thailand, Cambodia and other Southeast Asian countries. It was later introduced into Jamaica, Curaçao, and Aruba. It was introduced into the United States, into Florida in 1960.

Java apple blossom *(Prenn/Wikimedia Commons)*

TREE CHARACTERISTICS. A tropical evergreen tree, it grows from approximately 16 to 50 feet tall. It bears a wide canopy and a short, slightly crooked trunk 10 to 19 inches in diameter, with gray flaking bark. Its leaves are elliptical, measuring from four inches up to ten inches long and about two to five inches wide. The leaves are aromatic when bruised or crushed. New leaves are purplish blue but turn green as they mature.

The trees blossoms are white to yellowish white, with four petals measuring approximately an inch in diameter. The fragrant blossoms are borne in clusters of up to 30 on the tips of branches, axils of leaves, but can also appear on the surface of the trunk and branches. The blossoms are pollinated by insects. Mature trees can produce up to 700 Java apples a year. The fruit growing lower on the tree is said to be large, and the fruit growing higher in the canopy is supposedly smaller, but with greater color.

There are a variety of cultivars. Popular cultivars include the 'Black Pearl,' 'Jambu Madu Red,' 'Masam Manis Pink,' and the 'Pink Wax Jambu.' Trees can be propagated by seed, by budding, cuttings and by air layering. Trees grown from seed can take three to five years before beginning to bear fruit. However, trees propagated by air layering bear fruit in half that time.

► FAMILY: Myrtaceae. GENUS: *Syzygium* P. Br. Ex Gaertn. SPECIES: *Syzygium samarangense* (Blume) Merr. & L.M. Perry.

SELECTING A TREE. Different cultivars offer different levels of sweetness, acidity, and taste. Select a cultivar that best suits your taste. A frequently sought after cultivar is the 'Black Pearl' due to its taste.

When selecting a tree from the nursery, check the tree's overall health. Make sure its leaves are not discolored, and free from insects or signs of insect damage. Check the entire tree to make sure it is free from injury, such as broken branches, cuts in the trunk, etc. If possible, gently lift the tree out of the pot to check its roots. Avoid overgrown trees with girdled roots since they're more likely to develop a poor root structure.

HOW TO GROW A JAVA APPLE TREE. Java apple trees prefer warm moist climates above 65 degrees Fahrenheit. The climate is very important because the trees need temperatures consistently over 60 degrees to fruit. And abnormally high temperatures also thwart fruiting. The trees also prefer 59 to 91 inches of rainfall a year, but will tolerate rainfall between 47 and 118 inches per year.

The trees reportedly will grow in sandy soil, but prefer moist loamy fertile soils. If the tree is grown in poor soil the quality and quantity of the fruit declines. If you are growing a potted tree indoors, the trees will benefit from a regular application of a balanced fertilizer. Java trees prefer a soil pH of 6.0 to 7.5, but tolerate soil with a pH between 5 to 8. The trees grow best in full sun. To plant a tree, select a sunny location that receives a minimum of six hours of sun daily. Make sure the location you select can accommodate the size of a mature tree. Also try to select a location shielded from the wind since strong winds can damage the tree and bruise the fruit.

Once you've selected the best location, dig a hole at least twice the diameter of the pot the tree came in, about as deep as the container. Place the tree in the hole, loosening or cutting away any girdled roots. The tree should be planted so that the root ball is about one to two inches higher than the soil level. Refill the hole with the original soil. Water the newly planted tree thoroughly. The trees prefer moist, but not wet soil. Irregular irrigation is said to increase the risk of the fruit skin cracking.

The tree only needs minimal pruning. Pruning is only required to remove dead branches and to keep the canopy open to allow for sunlight. Pruning can also be done to maintain the tree size. Potential pests and diseases include oriental fruit flies, thirps and scales. The trees are also subject to sooty mold and bacterial wilt.

HOW TO GROW FROM SEED. Java apple seeds are only viable for a short period. So they should be planted shortly after they are removed from the fruit. Clean and wash away any pulp from the seed. Plant it in moist soil about half-inch deep. Keep the soil moist. It should germinate within six weeks. Note that trees grown from seed usually take three to five years before they begin fruiting.

► USDA HARDINESS ZONES: 10–11. CHILLING HOURS: 0. WATER REQUIREMENTS: More Than Average.

HOW TO CONSUME. Java apples are frequently eaten fresh. Make sure the apple is ripe. You can visually determine if the fruit is ripe because it will have a waxy appearance. Note that Java apples will not continue to ripen once they've been harvested. Java apples should be eaten shortly after harvest since they only store at room temperature for a couple of days, and are damaged if stored in the refrigerator. If you want to extend their shelf life, the University of Hawaii recommends storing them at 54 to 57 degrees Fahrenheit with 90 to 95 percent relative humidity. This should extend their shelf life to 10 to 14 days.

Java apples can be added to salads. If you want to preserve the fruit's unique shape, which makes a wonderful central focal point in a salad, you should remove the core without cutting the fruit into slices. Because of its high liquid content and mild flavor, it's often eaten fresh as a thirst quencher. It's also used to make a refreshing Java apple juice with sweetener and often other flavorings added. The fruit can be stewed and sautéed. Java apples can also be used in baked goods. You can find recipes for Java apple tarts and pies and other baked goods online.

ADDITIONAL INFORMATION. In some Asian countries the Java apple blossoms have been used to treat diarrhea and fevers. The tree wood has been used in construction.

REFERENCES

Al-Saif, Adel. M., et al. "Photosynthetic Yield, Fruit Ripening and Quality Characteristics of Cultivars of *Syzygium samarangense.*" *African Journal of Agricultural Research*, vol. 6, no. 15, August 4, 2011, p. 3623–3630.

Anza, Precy. "The Legend of Java Apple or Macopa." Hub Pages, August 8, 2018. https://hubpages.com/religion-philosophy/The-Legend-Of-Java-Apple. Accessed 10/26/18.

"BRIEF: Healthy Kitchen: Apples to Apples?" *Herald-Times* (Bloomington, IN), July 15, 2015.

Karp, David. "Wax Apples Come to Temple City, Pasadena Farmers Markets." *Los Angeles Times*, August 24, 2013. www.latimes.com/food/la-fo-market-news-online-20130824-story.html. Accessed 10/26/18.

Khandaker, Mohammad, et al. "Physiological and Biochemical Properties of Three Cultivars of Wax Apple (*Syzygium samarangense* (Blume) Merrill & L. M. Perry) Fruits." *Journal of Sustainability Science and Management*, vol. 10, no. 1, June 2015, p. 66–75.

Khandaker, Mohammad Moneruzzaman, and Amru Nasrulhaq Boyce. "Growth, Distribution and Physiochemical Properties of Wax Apple (*Syzygium samarangense*): A Review." *Australian Journal of Crop Science*, vol. 10, no. 12, 2016, p. 1640–1648.

Mattingly-Arthur, Megan. "Water Apples to Apples—Actually They're Berries." *San Diego Union Tribune*, August 20, 2014. www.sandiegouniontribune.com/news/health/sdut-water-apples-berries-waxy-2014aug20-story.html. Accessed 10/26/18.

McMullen, Samantha. "Jambu Fruit Plant Facts." SFGate. https://homeguides.sfgate.com/jambu-fruit-plant-104657.html. Accessed 10/19/18.

Morton, Julia F. *Fruits of Warm Climates*. Eugene, OR: Wipf and Stock Publishers, 2003.

"The Myth About the Macopa Fruit." Philippines Insider, January 25, 2018. www.philippinesinsider.com/myths-folklore-superstition/the-myth-about-the-macopa-fruit/. Accessed 10/26/18.

"The Origin of the Macopa Tree." Ginton Aral. www.gintongaral.com/myth-and-legends/the-origin-of-the-macopa-tree/. Accessed 10/26/18.

Orwa, C., et al. "*Syzygium samarangense*." 2009 Agroforestree Database: A Tree Reference and Selection Guide Version 4.0. http://www.worldagroforestry.org/treedb/AFTPDFS/Syzygium_samarangense.PDF. Accessed 03/29/16.

Paull, Robert E., and Ching Cheng Chen. *Wax Apple: Post Harvest Quality-Maintenance Guidelines*. F_N-39. College of Tropical Agriculture and Human Resources, University of Hawaii at Manoa, June 2014. www.ctahr.hawaii.edu/oc/freepubs/pdf/F_N-39.pdf. Accessed 11/03/18.

"Rose Apple Growing (Wax apple) Guide." AgriFarming. www.agrifarming.in/rose-apple-growing-wax-apple-guide/. Accessed 10/25/18.

"*Syzygium samarangense*." Encyclopedia of Life. http://eol.org/pages/2508668/details. Accessed 10/19/18.

"*Syzygium samarangense*." Flora Fauna Web, National Parks Service, Singapore. https://florafaunaweb.nparks.gov.sg/special-pages/plant-detail.aspx?id=4150. Accessed 10/19/18.

"*Syzygium samarangense*." National Tropical Botanical Garden. https://ntbg.org/database/plants/detail/syzygium-samarangense. Accessed 10/19/18.

"*Syzygium samarangense*." Tropical Plants Database. http://tropical.theferns.info/viewtropical.php?id=Syzygium+samarangense. Accessed 10/19/18.

"*Syzygium samarangense* (Blume) Merr. & L.M. Perry." USDA, NRCS. 2018. The PLANTS Database. National Plant Data Team, Greensboro, NC 27401-4901. https://plants.sc.egov.usda.gov/core/profile?symbol=SYSA3. Accessed 01/01/19.

Tuan, Nguyen Minh, and Yen Chung-Ruey. "Effect of Various Pollen Sources to Ability Fruit Set and Quality in 'Long Red B' Wax Apple." *International Journal of Biological, Biomolecular, Agricultural, Food and Biotechnological Engineering*, vol. 7, no. 2, 2013, p. 144–147.

"Wax Apple, Wax Jambu, Java Apple." Rare Fruit Club of Western Australia. www.rarefruitclub.org.au/Level2/WaxApple.htm. Accessed 10/19/18.

Java Plum (*Syzygium cumini* L. Skeels)

Also known as a jambolan, jamun, jambul, black plum, and Malabar plum, the fruit commonly referred to as a berry is oblong to round, usually measuring only half-inch to an inch or two long. When ripe the smooth thin skin turns from green to red to deep purple or black. The pulp is juicy and generally purple however white flesh cultivars exist. They generally have a single oblong seed, although seedless cultivars do exist. The purple pulp tends to color your tongue purple after consumption. The pulp is generally astringent with the flavor ranging from somewhat sweet to acidic. Its taste is often described and both sweet and sour at the same time.

JAVA PLUM LORE. It's said that Rama, the seventh avatar of the Hindi god Vishnu, survived by eating java plums during his 14-year exile from Ayodhya.

Because of that, some Hindus consider the java plum the "fruit of the gods." The java plum is considered sacred to Lord Krishna. He's often described as having the skin color of the Java plum. It's also said he had four symbols of the Java plum on his right foot. For this reason Java plum trees are often planted near Hindu temples because they're considered sacred.

TREE HISTORY. The Java plum is native to India, Burma, Ceylon and the Andaman Islands. It was introduced into Malaya and the Philippines, East Indies, and Queensland and New South Wales. The tree became established in Hawaii in 1870, where it is now considered an invasive species. The tree was officially introduced into the United States when the U.S. Department of Agriculture received seeds from the Philippines. In 1920s it was introduced into Puerto Rico, and Florida. In Florida the tree is also considered an invasive species. The tree is also grown in several countries, including but not limited to Bermuda, the French Islands of the Lesser Antilles and Trinidad, Polynesian Islands, African Continent, Australia, the Caribbean and South America.

TREE CHARACTERISTICS. Java plum trees are evergreen trees which can grow up to 100 feet tall in their native habitats. However, in the United States

Java plum *(Ton Rulkens/Wikimedia Commons)*

they generally only grow 40 to 50 feet tall. A fast growing tree, it has a dense canopy measuring up to 40 feet wide. It has a short, often crooked trunk 16 to 40 inches in diameter. A short distance from the ground the trunk usually branches into multiple trunks. Its bark is light gray and flaky lower on the trunk, and smoother higher up the tree. The trees leaves are elliptical, with an opposite leaf arrangement. The leaves measure from two to about ten inches long and up to four inches wide. New, young leaves start out pinkish in color then turn dark green as they mature, with a yellowish green midrib. The leaves have a slight turpentine scent.

The tree bears white blossoms a quarter- to half-inch wide. Each blossom has four petals. The blossoms are fragrant and appear in clusters. The blossoms are primarily pollinated by bees. The tree reportedly can live up to 100 or more years. It is propagated by seed, and by cuttings and grafting. Trees grown from seed usually begin bearing fruit in eight to ten years, whereas grafted trees usually being bearing fruit in four to seven years.

▶ FAMILY: Myrtaceae. GENUS: *Syzygium* P. Br. Ex Gaertn. SPECIES: *Syzygium cumini* (L.) Skeels.

SELECTING A TREE. When selecting a tree from the nursery, check the tree's overall health. Look for active growth, such as new or young leaves. Make sure its leaves are not yellow or discolored, and free from insects or signs of insect damage. Check the entire tree to make sure it is free from injury, such as broken branches, cuts in the trunk, etc. And if possible, gently lift the tree out of the pot to check the roots. Overgrown trees that have seriously girdled roots tend to develop a poor root structure so they should be avoided.

Cultivars include, but are not limited to 'Ram Jarnum,' which has purple pink pulp is juicy and sweet, and 'Krian Duat' with purple pulp.

HOW TO GROW A JAVA PLUM TREE. The tree is a tropical and subtropical tree. It grows in a wide variety of soils but fruits best in a well draining deep loamy soil. It prefers a soil pH of 5.5 to 7 and full sun. To plant a tree, select a sunny location that can accommodate the size of a mature tree. Dig a hole at least twice the diameter of the pot the tree came in, about as deep as the container. Place the tree in the hole, loosening or cutting away any girdled roots. The tree should be planted so that the top of the root ball is about one to two inches higher than the soil level. Refill the hole with the original soil. Water the newly planted tree thoroughly. Water the tree regularly, but do not over water it.

Pruning is only needed to remove dead branches and if you want to control the tree size. Reported tree pests include white flies, fruit flies, aphids and ants.

HOW TO GROW FROM SEED. Java plum trees propagate easily from seed. But the seeds are viable for only a short period, so they should be planted shortly after harvested. Remove the seed from the fruit, clean away any pulp and rinse the seed in water. Plant it in moist soil. Keep the soil moist, but not wet. The seeds will germinate in two to four weeks. Keep the soil moist, or the resulting seedlings will die.

▶ USDA HARDINESS ZONES: 9–13. CHILLING HOURS: 0. WATER REQUIRE-MENTS: Prefers 59–236 inches rainfall annually.

HOW TO CONSUME. Java plums have a very short shelf life. They can be stored at room temperature for only up to two days. The fruit can be eaten fresh. But because they are astringent, they are often soaked in salt water half an hour, to make them more palatable. The fruit is often used to make a beverage by washing the fruit, then cooking it in water for about a half hour. It's then strained and sugar is added.

Java plums are frequently made into wines and vinegars. The fruit is often made into a jelly. You can find java plum jelly recipes in the publication *Fruits of Hawaii: Description Nutritive Value and Use* (University of Hawaii Agricultural Experiment Station Bulletin #96) which is accessible on the web.

ADDITIONAL INFORMATION. The tree wood is hard and water resistant. It's used to make musical instruments, furniture, beams, etc. The tree bark is used in dyes and tanning leather. In folk medicine the seeds and leaves are used to treat diabetes and other health conditions.

REFERENCES

Binita, Kumari, et al. "The Therapeutic Potential of *Syzygium cumini* Seeds in Diabetes Mellitus." *Journal of Medicinal Plants Studies*, vol. 5, no. 1, 2017, p. 212–218. www.plantsjournal.com/archives/2017/vol5issue1/PartC/5-1-40-715.pdf. Accessed 10/29/18.

Jadhav, V.M., S.S. Kamble, and V.J. Kadam. "Herbal Medicine: *Syzygium cumini* :A Review." *Journal of Pharmacy Research*, vol. 2, no. 8, 2009, p. 1212–1219. http://jprsolutions.info/files/final-file-56b225bf628a11.26562478.pdf. Accessed 10/28/18.

"Jambolan—*Syzygium cumini* (L.) Skeels." Growables.org www.growables.org/information/TropicalFruit/Jambolan.htm. Accessed 10/29/18.

"Jamun (*Syzygium cumini*)." Horticultural College and Research Institute, Tamil Nadu Agricultural University, Coimbatore, India. http://agritech.tnau.ac.in/horticulture/horti_fruits_jamun.html. Accessed 10/29/18.

Miller, Carey D., and Katherine Bazore. *Fruits of Hawaii: Description, Nutritive Value and Use*. University of Hawaii Agricultural Experiment Station Bulletin no. 96. University of Hawaii, Honolulu. 1945. www.ctahr.hawaii.edu/oc/freepubs/pdf/B-96.pdf. Accessed 11/03/18.

Morton, Julia F. *Fruits of Warm Climates*. Eugene, OR: Wipf and Stock Publishers, 2003.

Morton, Julia F. "The Jambolan (*Syzygium Cumini* Skeels)—Its Food, Medicinal, Ornamental and Other Uses." Proceedings of the Florida State Historical Society. 1963. p. 328–338. https://fshs.org/proceedings-o/1963-vol-76/328-338%20(MORTON).pdf. Accessed 11/01/18.

Orwa, C., et al. "Syzygium cumini." 2009 Agroforestree Database: A Tree Reference and

Selection Guide Version 4.0. www.worldagroforestry.org/treedb/AFTPDFS/Syzygium_cuminii.PDF. Accessed 10/29/18.

Shukla, Rakesh. "JAMBU (Syzygium cumini)—A FRUIT OF GODS." *Ayurpharm International Journal of Ayurveda and Allied Sciences*, vol. 2, no. 4, 2013, p. 86–91. https://www.researchgate.net/publication/313741569_JAMBU_Syzygium_cumini_-A_FRUIT_OF_GODS/. Accessed 10/29/18.

Sivasubramaniam, K., and K. Selvarani. "Viability and Vigor of Jamun (Syzygium cumini) Seeds." *Brazilian Journal of Botany*, vol. 35, no. 4, 2012, p. 397–400.

Swami, Shrikant Baslingappa, et al. "Jamun (*Syzygium cumini* (L.)): A Review of Its Food and Medicinal Uses." *Food and Nutrition Sciences*, vol. 3, 2012, p. 1100–1117.

"*Syzygium cumini.*" Encyclopedia of Life. http://eol.org/pages/2508660/details. Accessed 10/29/18.

"*Syzygium cumini.*" Flora of the Hawaiian Islands. Botany Department, Smithsonian Institution. http://botany.si.edu/pacificislandbiodiversity/hawaiianflora/speciesdescr.cfm?genus=Syzygium&species=cumini. Accessed 10/29/18.

"*Syzygium cumini.*" Florida Invasive Plant Species Mobile Guide. www.plantatlas.usf.edu/flip/plant.aspx?id=25. Accessed 10/29/18.

"*Syzygium cumini.*" Grow Plants. www.growplants.org/growing/syzygium-cumini. Accessed 10/29/18.

"Syzygium cumini—(L.) Skeels." Plants for a Future. https://pfaf.org/user/Plant.aspx?LatinName=Syzygium+cumini. Accessed 10/29/18.

"Syzygium cumini—(L.) Skeels." Tropical Plants Database, Ken Fern. http://tropical.theferns.info/viewtropical.php?id=Syzygium+cumini. Accessed 10/29/18.

"Syzygium cumini—(L.) Skeels." USDA, NRCS. 2018. The PLANTS Database. National Plant Data Team, Greensboro, NC 27401-4901. https://plants.usda.gov/core/profile?symbol=SYCU. Accessed 10/29/18.

Jujube (*Ziziphus jujuba* [L.] Karst)

Also known as a Chinese date or a red date, or a Chinese plum, the jujube is usually an oval shaped (although some cultivars are round) white fleshed fruit with a thin edible dark reddish brown skin when ripe. The fruit generally ranges in size from half- to one-inch, and contains one pit in the center.

Many cultivars of jujubes exist. Those cultivars meant to be dried or otherwise processed tend to be rather dry and not very tasty. When dried, those cultivars taste like dates. Whereas newer cultivars intended to be eaten fresh are sweeter and juicier. When eaten fresh, these cultivars taste similar to a blend of an apple and a date.

The fruit starts out with a green skin that turns yellowish with brown blotches as it begins to mature. When ripe, the skin turns reddish brown and will begin to wrinkle and the flesh begins to soften. Green jujubes do not ripen once picked from the tree. Jujubes picked when their skin is yellowish can be ripened at room temperature. But those jujubes allowed to fully ripen on the tree will have the highest sugar content.

JUJUBE LORE. Jujubes are said to be a "lucky" fruit in Asian cultures. It's frequently eaten at the start of a new lunar year to bring good fortune. According to legend, the smell of jujubes has the power to make people fall in love.

Jujube *(www.maxpixel.com)*

In the Himalayas single individuals has been known to wear jujube blossoms to attract the opposite sex. And after a wedding, jujubes are often placed in the newlyweds' bedroom to bring good luck in fertility.

In ancient floral vocabulary, the jujube tree is a symbol for relief. Jujubes have been used in Chinese medicine to treat a variety of conditions. Jujubes are used to treat diabetes, liver disease, anemia, high blood pressure, anxiety, insomnia, inflammation, ulcers and muscular conditions. But there's currently insufficient evidence that jujubes are effective for treatment of these conditions.

TREE HISTORY. Originating in China, the tree was introduced into Europe at the beginning of the Christian era. The fruit also made its way to, and is grown in, the Middle East, Africa and Russia. It made its way from Europe to the United States, specifically North Carolina in 1837. It was later introduced into California, and then Gulf Coast States. But it wasn't until 1908 that several commercial jujube cultivars were brought into the USDA Plant Introduction Station in Chico (CA) by Frank Meyer. Unknown at the time was that the majority of these early imports were meant to be dried, not eaten fresh. In subsequent years cultivars meant to be eaten fresh were imported and spurred the popularity of the fruit in the U.S.

Dried jujubes *(www.maxpixel.net)*

Sixty to 70 jujube cultivars can be found in the United States, although there are over 800 cultivars in China, which has actively bred new cultivars over the years. Sample popular cultivars include the following: 'Li'—produces big round fruit up to two inches in diameter, sweet, good for eating fresh. • 'Lang'—produces oblong to pear shaped fruit up to 2 inches long, good eaten dried. • 'Honey Jar'—produces small sweet fruit good for eating fresh. • 'Shui Men'—produces jujubes that can be enjoyed eaten both fresh and dried. • 'Redland'—produces large fruit.

TREE CHARACTERISTICS. A deciduous tree, they generally grow 15 to 30 feet tall, but in more favorable climates like Florida have grown up to 40 feet tall. Depending upon the cultivar, the tree canopy could be broad, or narrow and upright. Hence the canopy can be 10 to 30 feet wide. The ovate shaped leaves are shiny green and one to two inches in length. The leaf stems are frequently described as zigzagged with two spines at the base of each leaf. In the autumn the leaves turn yellow.

The tree bears tiny (generally less than quarter-inch) yellow flowers with a green tint. The flower blossoms are fragrant. Depending upon the cultivar, the blossoms may be singular or in clusters. Although some cultivars are self-fertile, cross pollination increases the fruit production. Cross pollination can occur from the wind, or by insects.

Most jujube trees have thorns. But cultivars without thorns do exist. The tree wood is brown and hard. Most trees begin to fruit in their second year. The trees have a long life and continue to produce fruit. Some trees in China are believed to be 1000 years old.

The trees tolerate cold weather. There have been reports that they are hardy to anywhere from minus 10 degrees Fahrenheit to minus 28 degrees Fahrenheit. In the United States the jujube trees are fairly disease and pest free. Whereas in China, the primary disease is Witches Broom, which is a bacterial disease, and *Carposina niponensiss*, a moth, is their primary pest.

▶ FAMILY: Rhamnaceae. GENUS: *Ziziphus* Mill. SPECIES: *Ziziphus jujube* (L.) Karst.

SELECTING A TREE. When selecting a tree, be sure to select a cultivar that satisfies your desired use. If your goal for planting a tree is to have fresh jujubes to eat, you many want to consider selecting a 'Li' or 'Honey Jar' cultivar. If your goal is to have jujubes to dry, consider a cultivar like the 'Lang.' Or if you want jujubes that are good for both eating fresh, and to be dried, consider cultivars like the 'So' and 'Shuimen.'

Jujubes trees can be grown from seed. But trees grown from seed are not true to type. So it's better to select a tree that was propagated by grafting or budding.

Most jujube cultivars have thorny branches. If you are not a fan of thorny branches, or have young children, you may want to consider a thornless cultivar.

When selecting a tree from the nursery, check the tree's overall health. Make sure its leaves are not discolored, and free from insects or signs of insect damage. Check the entire tree to make sure it is free from injury, such as broken branches, cuts in the trunk, etc. Check to make sure the tree trunk is not sunburned. If possible, lift the tree out of the pot to check the roots. If the tree has significant girdled roots it should be avoided because it is likely to develop a poor root structure.

GROWING JUJUBES. Jujubes grow best in full sun. Besides selecting a sunny location to plant your tree, be sure the location selected can accommodate the mature size of the cultivar selected. Also bear in mind that mature trees can produce about 100 pounds of jujubes. So select a location suitable for fruit drop from unharvested jujubes. If you are planting more than one tree, they should be spaced at least 15 feet from one another.

Jujubes grow in most soils provided there is adequate drainage. When planting the tree, dig a hole twice the diameter of the tree container and as deep as the container. Place the tree in the hole so that the top of its root ball sits an inch or two above the soil line, and backfill with the original soil. After planting the tree, water it thoroughly. Although the trees are drought tolerant, newly planted trees should be watered regularly until they become established.

Do not fertilize newly planted trees. There's a lack of sufficient research on fertilization on jujube trees for home gardens. Jujube trees do not require fertilization for growth. However trees grown commercially tend to be fertilized by their growers. Like other fruit trees, jujube trees should be pruned annually to remove any dead wood, and to maintain your desired tree size. Pruning also allows more sunlight to reach various areas of the tree.

How to Grow from Seed. Growing a jujube tree from seed is generally not preferred because of the low germination rates, and the tree may not be true to type. But if you hate to see seeds go to waste and want to test your green thumb, here's how to grow a jujube tree from seed.

First, wash away any fruit pulp from the pits. If you don't plan on immediately planting the seeds, seal them in a plastic bag and don't allow them to dry out. Otherwise scarify the seeds by pouring really hot/boiling water over the seeds and let it soak for 24 hours. (Scarification is done to weaken the seed coat and encourage germination.)

Next stratify the seeds by keeping them in a dark place at least 68 degrees Fahrenheit for three months. Then keep the seeds in a cool place (40 degrees Fahrenheit) like the refrigerator for three months. Plant the seeds half an inch deep in the soil in really warm temperatures. Keep the soil most. The seeds should germinate in about four weeks. If your seeds don't sprout, just remember that jujube seeds have a low germination rate.

▶ USDA Hardiness Zones: 6–10. Chilling Hours: 150–200. Water Requirements: Average.

How to Consume. To eat them fresh, pick jujubes from the tree whose skin has turned dark red/brown and has begin to crinkle. The pulp should be soft. Some people like to eat fresh jujubes when the skin has begun to turn dark, but the flesh is still crisp like an apple. Note that jujubes picked green do not ripen satisfactorily at room temperature and have a lower sugar content. Fresh jujubes can be stored for up to 70 days if refrigerated at 50 degrees Fahrenheit. If refrigerated at below 36 degrees Fahrenheit, large sunken areas on the jujubes' skin will appear. The jujubes are frequently dried. Jujubes can be sun dried (not recommended), dried in an oven, and dried in an electric dehydrator.

For food safety reasons, sun dried jujubes need to be pasteurized to kill any insects or their eggs. You can pasteurize them by placing the dried jujubes in a sealed freezer plastic bag and letting them sit in the freezer (must be at zero degrees Fahrenheit) for at least 48 hours. Or you can pasteurize them in your oven by placing them on a shallow tray in a single layer. Heat them for 30 minutes at 160 degrees Fahrenheit.

Information on how to dry jujubes can be found on the National Center for Home Food Preservation website at www.nchfp.uga.edu. Dried jujubes can be used to replace raisins and dried dates in baked goods. Jujubes are often candied, made into syrup, and made into a filling for pastries and other baked goods. They are also used in jams. You can find selected jujube recipes on the California Rare Fruit Growers Jujube page on their website at www.crfg.org/pubs/ff/jujube.html.

ADDITIONAL INFORMATION. Jujube extracts are used in the manufacturing of skin care products for relief of dry skin, to diminish wrinkles, and for sunburn relief. Jujubes contain chemicals that act like antioxidants. Jujubes are very high in vitamin C, with .22 pounds of fresh jujubes providing more than 100 percent of the recommended daily value of vitamin C.

Jujubes have been made into wine and have been mixed with hard liquor and made into a spirit. Jujubes are also used to make tea, and tea syrup. China produces 90 percent of the world's production of jujubes.

REFERENCES

Ashton, Richard. "Jujube: A Fruit Well Adapted to Texas." *Texas Gardener.* January/February 2008. www.texasgardener.com/pastissues/janfeb08/Jujube.html. Accessed 02/22/18.

Folkard, Richard. *Myths, Traditions, Superstitions and Folk Lore of the Plant Kingdom.* London: Sampson, Low, Marston, Searle and Rivington. 1884.

Gilman, Edward F., and Dennis G. Watson. *Ziziphus jujuba: Chinese Date.* ENH-836. University of Florida, IFAS Extension. 2006. http://edis.ifas.ufl.edu/st680. Accessed 02/14/16.

"Jujube." California Rare Fruit Growers. 1996. www.crfg.org/pubs/ff/jujube.html. Accessed 02/22/2015.

"*Jujube and Aronia* (Crop Profile)." Center for Crop Diversification, University of Kentucky. University of Kentucky, Cooperative Extension Service. Feb. 2013. www.uky.edu.ccd/files/jujube.pdf. Accessed 02/23/18.

"*Jujube (Ziziphus Jujuba).*" United Nations Institute for Training and Research. Green Legacy Hiroshima. www.unitar.org/hiroshima/sites/unitar.org.hiroshima/files/GLH%202011%20Tree%20Jujube_0.pdf. Accessed 03/03/18.

"Jujubes." National Gardening Association Plants Database. https://garden.org/plants/group/jujubes/. Accessed 01/09/19.

"Jujubes Fresh Eating." New Mexico State University, College of Agricultural, Consumer, and Environmental Sciences. https://aces.nmsu.edu/jujube/eating.html. Accessed 01/09/19.

"Jujubes in Western Australia—A Growing Industry." WA Jujube Growers Association. www.agric.wa.gov.au/sites/gateway/files/Jujube%20Industry%20prospectus.pdf. Accessed 03/04/18.

Kader, Adel A., Alexander Chordas, and Yu Li. "Harvest and Postharvest Handling of Chinese Date." *California Agriculture.* January/February 1984. p. 8–9.

Martin, Chris. *Ziziphus Jujuba.* Virtual Library of Phoenix Landscape Plants. Arizona State University. www.public.asu.edu/~camartin/plants/Plant%20html%20files/ziziphusjujube.html. Accessed 02/14/16.

Plastina, Pierluigi. "Pharmacological Aspects of Jujubes." *Pharmacologia*. v. 7, p. 243–255. June 15, 2016.

"SelecTree: *Ziziphus jujuba* Tree Record." California Polytechnic State University, Urban Forest Ecosystems Institute. 1995–2018. http://selectree.calpoly.edu/tree-detail/ziziphus-jujuba. Accessed 03/03/18.

Yao, Shengrui. *Jujube: Chinese Date in New Mexico, Guide H-330*. New Mexico State University, Cooperative Extension Service, May 2012. https://aces.nmsu.edu/pubs/_h/H330/. Accessed 04/02/18.

Yao, Shengrui. "Past, Present and Future of Jujubes—Chinese Dates in the United States." *HortScience*, v. 48, no. 6, June 2013, p. 672–680.

"Ziziphus." WebMd. www.webmd.com/vitamins-supplements/ingredientmono-62-zizyphus. aspx?activeingredientid=62&activeingredientname=zizyphus. Accessed 02/26/18.

"Ziziphus Jujube Seeds Chinese Ziziphus Tree." Zhong Wei Horticultural Products Company. www.cnseed.org/ziziphus-jujube-seeds-chinese-ziziphus-tree.html. Accessed 03/03/18.

Kaffir Lime (*Citrus hystrix* DC)

The Kaffir lime is also known as a k-lime, Keiffer lime, a Makrut lime, a Magrut lime, and a Thai lime. (Because the name Kaffir is a derogatory term, it is being referred to as a k-lime in the rest of this section. For more information, see the Additional Information section of this chapter.) The k-lime is a small green lime easily distinguishable by its lumpy exterior. The lime's peel is dark green when immature, and turns yellow when mature. Measuring only about two inches in diameter, each lime has several seeds. The pulp's flavor is described as bitter or tart, with a unique taste something like a combination of a lemon, lime and orange. The lime doesn't have much juice. The lime's peel and the plant's leaves are used more than the fruit's pulp and juice. The k-lime's juice is frequently described as unpalatable.

K-LIME LORE. In Malaysia there's a superstition that the k-lime wards off evil spirits. In Chinese mythology, during the new lunar year it's auspicious to bathe in water containing lime juice. In Thai folklore the k-lime is said to cleanse the mind and body. In some Southeast Asian countries the k-lime's juice is believed to kill head lice when used as a shampoo. It's also believed to kill leeches that have attacked the feet when used as a foot wash. In Indonesia, the limes are used for gum and dental health.

TREE HISTORY. The tree originated in Indochina and the Malesian region. In addition to Asia, it's now also grown in suitable climates around the world, including Central America, Africa, Australia, and the United States.

TREE CHARACTERISTICS. An evergreen tree, it generally grows five to ten feet tall (although reportedly can grow up to 35 feet tall in ideal conditions). They often appear more like a shrub than a small tree. The tree has thorny

Kaffir lime *(Katja Schulz from Washington, D.C., USA/Wikimedia Commons)*

multi-branches. Individual thorns can be up to two cm long. Its leaves are distinctive shiny dark green "double leaves" that are shaped like hour glasses. The leaves are extremely fragrant and used in Asian cooking. The tree bears small fragrant white blossoms with four to five petals with slight pink on the outside. The tree is self fruitful. Bees help with pollination. The tree can be propagated by seed and vegetatively (grating, cuttings, etc.). Grafted trees tend to bear limes in their second year. However trees propagated by seed can take up to ten years before bearing fruit.

▶ FAMILY: Rutaceae. GENUS: *Citrus* L. SPECIES: *Citrus hystrix* DC.

SELECTING A TREE. When selecting a tree from the nursery, check the tree's overall health. Look for active growth, such as new or young leaves. Make sure its leaves are not yellow or discolored, and free from insects or signs of insect damage. Check the entire tree to make sure it is free from injury, such as broken branches, cuts in the trunk, etc. And if possible, gently lift the tree out of the pot to check the roots. Overgrown trees that have seriously girdled roots will likely develop a poor root structure so they should be avoided.

HOW TO GROW A K-LIME TREE. The tree prefers full sun, a humid environment, and ideally a soil pH between 6.0 to 6.5. The best time to plant a lime tree is in the spring when there's no longer any danger of frost. Planting in the spring allows the tree to become established before the winter.

To plant a tree, select a sunny location or one with partial shade that can accommodate the size of a mature tree. Dig a hole at least twice the diameter of the pot the tree came in, about as deep as the container. Place the tree in the hole, loosening or cutting away any girdled roots. The tree should be planted so that the top of the root ball is about one to two inches higher than the soil level. Refill the hole with the original soil. Water the newly planted tree thoroughly. The tree should be regularly watered the first two years.

Young citrus trees are sensitive to frost. If temperatures are expected to dip to 29 degrees Fahrenheit or lower, you need to protect the tree from frost. You should wrap the tree trunk and branches in insulation material like cardboard, fiberglass or frost protection covers. Make sure the soil is moist and not dry, since damp soil retains and radiates more heat than dry soil. Bare soil also radiates more heat than soil covered with mulch or other ground covers.

If your tree does sustain frost damage, don't prune away any dead branches until the spring. This allows time for the tree to recover in warmer weather. It also allows you to better identify the damaged branches to be removed. Young citrus trees benefit from fertilization. If using a citrus fertilizer, follow the instructions on the label.

HOW TO GROW FROM SEED. To grow a tree from seed, remove the seeds from the lime. Wash any pulp away from the seeds. Soak the seeds in clean water overnight. Place the seeds about half-inch deep in a sterile moist planting/potting medium. Cover the pot with plastic to create a humid atmosphere. Keep the soil moist until seedlings sprout.

The seeds germinate best in temperatures above 60 degrees Fahrenheit. The seeds usually germinate in two to three weeks, but can take up to three months. Once the seeds sprout, transplant them into a larger pot and move them to a sheltered location outdoors. Be aware that trees grown from seed do not always bear true to type. And trees grown from seed may take up to ten years before bearing fruit.

▶ USDA HARDINESS ZONES: 9–11. CHILLING HOURS: 0. WATER REQUIREMENTS: Average.

HOW TO CONSUME. Before using k-lime peel or leaves, be sure to wash them thoroughly. K-lime leaves (both fresh and dried) and peels are frequently used in Thai cooking. The addition of grated k-lime rind to Thai curries is what distinguishes it from Indian curries. Whole fresh or dried k-lime leaves are added to boiling mixtures when making Thai soup stock since they are ultimately removed after cooking. Whole fresh k-lime leaf is also added to the water used to cook rice or noodles for flavoring, then removed once the rice or noodles are cooked. The lime leaf flavoring is a staple in making *tom yum* soup. Hair-like slivers of fresh k-lime leaves are sometimes added to fresh salads for flavor.

Fresh limes should only be stored at room temperature for less than seven days. For longer storage, they should be placed in the refrigerator.

K-lime peel can also be dried for later use. To dry the peel, wash the peel thoroughly in water. Dry it with a paper towel. Then remove the outer sixteenth to quarter-inch of peel, avoiding the pith. Place the peel pieces single layer cut side up on your electric dehydrator tray. Set the temperature at 130 to 135 degrees Fahrenheit. The estimated drying time is eight to 12 hours. You should dehydrate the peels only until they are dry, careful not to over dry them. To test them to see if they are dry, remove a couple of pieces from the dehydrator and let them cool off. Then slice them in half. You should not see any moisture in them, nor should you be able to squeeze any moisture out of them. The peel should still be pliable, but not sticky, which means if you fold it in half it should not crack, nor should it stick to itself.

When the peel is dry, remove it from the dehydrator and let it cool for an hour. Then place the dried peel in an airtight container, like a canning jar or plastic freezer bag, marking the date on the container, and store it in a dark cool (60 degrees Fahrenheit) place. Use the peel within a year, and be sure to discard any moldy or spoiled dried peel.

A study in the *Journal of Food, Agriculture and Environment* showed that the optimum hot air drying time for k-lime leaves was 4.9 hours at 140 degrees Fahrenheit, and a loading capacity 1.4 kg per square meter.

ADDITIONAL INFORMATION. WARNING: If you are taking a medication that is broken down by your liver, you should consult your doctor before consuming lime juice. Lime juice could decrease how fast your liver breaks down certain medications, increasing its effect and any side effects. There are many medications that are changed by your liver. Lovastatin and fexofenadine are just two of many changed by the liver.

NOTE: The word "kaffir" is considered a racist term. The word originated from the Arabic word kafir meaning non-believer or infidel. The word was misinterpreted by Portuguese slave traders and over time evolved into a derogatory hate word against blacks. The word is actionable in court in South Africa. There is an effort underway to rename the lime simply a k-lime.

K-lime juice is used to clean clothes in Thailand, and the leaves are used as an insect repelling shampoo due to them containing citronellal. A study conducted on rabbits showed that k-limes helped fight bacteria like e. coli.

REFERENCES

Andress, Elizabeth L., and Judy A. Harrison. *So Easy to Preserve*. 6th ed. Bulletin 989. Athens, GA: College of Family and Consumer Sciences, College of Agricultural and Environmental Sciences, Cooperative Extension, University of Georgia. 2014.

"Citrus for Hawai'i's Yards and Gardens." F&N-14. College of Tropical Agriculture and Human Resources, University of Hawai'i at Manoa. June 2008. www.ctahr.hawaii.edu/oc/freepubs/pdf/F_N-14.pdf. Accessed 8/03/18.

"*Citrus hystrix.*" Encyclopedia of Life. http://eol.org/pages/2906075/details. Accessed 08/03/18.

"*Citrus hystrix.*" Missouri Botanical Garden. www.missouribotanicalgarden.org/PlantFinder/PlantFinderDetails.aspx?taxonid=291778&isprofile=0&. Accessed 08/03/18.

"*Citrus hystrix DC.*" USDA, NRCS. 2018. The PLANTS Database. National Plant Data Team, Greensboro, NC 27401-4901. https://plants.sc.egov.usda.gov/core/profile?symbol=CIHY2. Accessed 07/31/18.

"Citrus—Makrut Lime, Kaffir." LEAF Network, Az. www.leafnetwork.org/resources/Citrus_Lime_kaffir_profile.pdf. Accessed 07/13/18.

Cunningham, Sally. *Asian Vegetables: A Guide to Growing Fruit, Vegetables and Spices from the Indian Subcontinent*. Bath (UK): Eco-Logic Books. 2009.

Fanous, Summer. "8 Healthy Lime Facts." *Healthline Newsletter*. May 5, 2016. https://www.healthline.com/health/8-healthy-lime-facts. Accessed 08/01/18.

Geisel, Pamela M., and Carolyn L. Unruh. *Frost Protection for Citrus and Other Subtropicals*. ANR Publication 8100. University of California, Division of Agriculture and Natural Resources, 2003.

Juhari, Nurul Hanisah, et al. "Effect of Hot-air Drying on the Physicochemical Properties of Kaffir Lime Leaves." *Journal of Food, Agriculture & Environment*, vol. 11, no. 1, January 2013, p. 203–211.

Kader, Adel, et al. *Storing Fresh Fruits and Vegetables for Better Taste*. Postharvest Technology Center, University of California, Davis. 2012. http://ucce.ucdavis.edu/files/datastore/234-1920.pdf. Accessed 09/10/18.

"Kaffir Lime." Citrus Variety Collection, University of California, Riverside. www.citrusvariety.ucr.edu/citrus/hystrix_2454.html. Accessed 07/30/18.

"Kaffir Lime." The Epicentre. http://theepicentre.com/spice/kaffir-lime/. Accessed 08/02/18.

"Kaffir Lime." www.doc-developpement-durable.org/file/Arbres.../Kaffir%20lime.pdf. Accessed 08/03/18.

"Kaffir Limes." Specialty Produce. www.specialtyproduce.com/produce/Kaffir_Limes_2163.php. Accessed 07/30/18.

Keah, SH, and KS Chng. "A Chinese New Year Rash." *Malaysian Family Physician*, vol. 8, no. 2, 2013, p. 62–64.

LeBlanc, Tyler. "Why the Name 'Kaffir Lime' Is Wildly Offensive to Many." *Modern Farmer*, July 2014. https://modernfarmer.com/2014/07/getting-rid-k-word/. Accessed 07/30/18.

Loha-unchit, Kasma. "Kaffir Lime—Magrood." Thai Food and Travel. www.thaifoodandtravel.com/ingredients/klime.html. Accessed 08/02/18.

Long, Jim. "Herb to Know: Kaffir Lime." *Mother Earth Living*, April/May 2005. www.motherearthliving.com/plant-profile/herb-to-know-kaffir-lime. Accessed 07/30/18.

Mullins, Lynne. "Kaffir Limes and Leaves." *Sydney Morning Herald*, September 2, 2008, Good Living Section, p. 15.

"9 Surprising Benefits of Kaffir Lime." www.organicfacts.net/health-benefits/fruit/kaffir-lime.html. Accessed 07/30/18.

Ortho All About Citrus and Subtropical Fruits. Des Moines, IA: Meredith Books, 2008.

Piper, Jacqueline M. *Fruits of South-East Asia: Facts and Fiction*. Oxford, NY: Oxford University Press, 1989.

Sacks, Katherine. "Why You Should Zest All Your Citrus." *Epicurious*, June 10, 2015. www.epicurious.com/expert-advice/why-you-should-zest-all-your-citrus-article. Accessed 08/03/18.

Snart, Jennifer E., Mary Lu Arpaia, and Linda J. Harris. *Oranges: Safe Methods to Store, Preserve and Enjoy*. Publication 8199. University of California, Division of Agriculture and Natural Resources, 2006. http://homeorchard.ucdavis.edu/8199.pdf. Accessed 08/03/18.

Sterman, Nan. "Grow Your Own Exotic Citrus—Pummelos, Blood Oranges and Kaffir Limes."

Los Angeles Times, California Living Section, July 3, 2008. www.latimes.com/style/la-hm-citrus3-2008jul03-story.html. Accessed 08/03/18.
Virgil, Evetts. "Kaffir Lime Leaf Gives Heady Hit." *The Timaru Herald*, July 9, 2013, p. 10.

Kiwifruit (*Actinidia spp.*)

In Chinese the fruit's original name is *mihoutao*. In English it was originally known as Chinese gooseberries (although botanically speaking it is not a gooseberry), kiwifruit, also known as just kiwi, are oblong fruits usually measuring about two inches long. The most common kiwi species has russet-brown skin covered with short stiff brown hairs. The skin is inedible; however some species feature edible skins. The pulp is green with a whitish center and tiny edible black seeds. The pulp is firm until it ripens. One kiwifruit species has a "hairless" skin, and some cultivars have yellow pulp. Kiwifruit pulp is juicy. They taste like a cross between a gooseberry and a strawberry. They are considered subacid to acidic.

KIWI LORE. In folk medicine kiwifruit has been used to treat everything from rheumatoid arthritis, to hepatitis, cancer, high blood pressure and asthma. However, at this time there's a lack of sufficient studies to support the effectiveness of kiwifruit for these conditions.

PLANT HISTORY. Native to the Yangtze River area of Northern China, kiwifruit plants date back to at least the 18th century. Kiwifruit seeds were sent to England in 1900, with plants resulting from the seeds. In 1904 the U.S. Department of Agriculture received kiwifruit seeds which successfully grew plants that bore fruit at the Plant Introduction Station that existed then in Chico, California. Seeds were then introduced into New Zealand in 1906 where the fruit thrived into a booming industry. Kiwi budwood was later introduced into South Africa in 1960.

Although the fruit originated in China, its original English name of Chinese gooseberries was changed in 1959 to its now popular name by New Zealand exporters Turners and Growers. They changed, some say hijacked, the name because gooseberries are not that popular and the fruit needed a new name to be commercially viable. So they renamed the fruit kiwi after their furry brown flightless national bird about the size of a chicken. Then in the 1980s, New Zealand plant physiologist Ross Ferguson reclassified the popular fuzzy skinned green pulped fruit from *Actinida chinensis* to *Actinida deliciosa*.

PLANT CHARACTERISTICS. There are approximately 50 species of *Actinidia*, but only a handful of species produce popular edible kiwifruit. The most

Kiwi *(www.maxpixel.com)*

popular species is *Actinidia deliciosa*, which produces the green pulped fuzzy skinned kiwifruit most commonly sold in supermarkets. Other species include the *Actinidia chinensis*, which bears a yellowish or golden pulped kiwifruit, *the Actinidia arguta*, which bears a small, berry sized fruit with a smooth green edible skin and is also known as the hardy Kiwi because it can tolerate cold temperatures as low as minus 25 degrees Fahrenheit, and *Actinidia kolomikta*, also known as Arctic kiwi, which is another cold-hardy species that can withstand very cold temperatures (often to minus 40 degrees Fahrenheit) and produces small fruits the size of a grape.

The most popular kiwifruit plants, *Actinidia deliciosa*, are considered subtropical plants. It's considered a fast growing woody vine plant that can grow up to 12 feet high, and its vines generally grow 10 to 15 feet or longer. The vines bear large, dark green oval leaves 7 to 10 inches in diameter. The plant's blossoms are large (one to two inches wide), fragrant, and start out white then turn cream colored.

The plants are female or male. (Although a few new cultivars are self fertile.) So generally you will need to plant both a male plant and female plant if you want them to bear fruit. The blossoms are pollinated by bees, other insects and birds. Although the plants can be propagated by seed, propagation is mainly done by grafting and softwood cuttings. Plants grown from seed are usually grafted at one year of age so that their sex is known.

▶ FAMILY: Actinidiaceae. GENUS: *Actinidia* Lindl. SPECIES: *Actinidia deliciosa*; *Actinidia chinensis*; *Actinidia arguta*; *Actinidia kolomikta*.

Kiwi blossom *(Rob Hille/Wikimedia Commons)*

SELECTING A PLANT. Successfully growing a kiwifruit plant starts with selecting a species and cultivar suitable for your climate and paying attention to the chilling hours needed for the cultivar. But taking the time to research a good plant to grow in your region will pay off with delicious kiwifruit for your consumption. But while doing your research, don't forget to select a cultivar that you like to eat! Cultivars in species *Actinidia deliciosa and Actinidia chinensis* are hardy in USDA zones 5 to 9. These species need approximately 225–240 frost-free days for the fruit to ripen. Some cultivars in species *Actinidia arguta* and *Actinidia kolomikta* are hardy down to USDA zone 3. *Actinidia arguta* plants need approximately 150 frost-free days for the fruit to ripen, and *Actinidia kolomikta* needs approximately 130 frost-free days.

You really need to make sure the cultivar you select is a good cultivar for your specific region. Below are some sample cultivars in each species.

Actinidia deliciosa cultivars—'Hayward'—fruit from this cultivar is most often found in supermarkets. • 'Meander'—another common cultivar. • 'Saanichton 12'—hardier than the Hayward but the pulp is supposedly tough and not soft. • 'Bruno'—a darker cultivar than the Hayward, but it's noteworthy because it only requires 50–250 chilling hours. Other low chilling hour cultivars include the 'Abbott,' 'Allison,' 'Blake,' 'Elmwood,' 'Tewi' and 'Vincent.'

Actinidia chinesis cultivars—'Hortl6A' is sold under the name Zespri Gold Kiwifruit. • 'Lushanxiang' is a cultivar grown in California. • 'Golden Yellow' is another cultivar grown in California. • 'Golden Sunshine' is a cultivar needing 700 chilling hours for bud break. • 'Golden Dragon' needs 800 chilling hours for bud break.

Actinidia arguta cultivars—'Issai' is a self fertile cultivar, however it will produce larger fruit when cross-pollinated. • 'Dumbarton Oaks' is said to be a highly productive cultivar. • 'Geneva' is a cultivar known for ripening earlier than the Issai. • 'Ananasnaya' is a popular hardy kiwi, which has green to purplish-red skin. (Note there is also another cultivar in the *Actinidia kolomikta* species with the same name.)

Actinidia kolomikta cultivars—'Pautske' is a cultivar said to produce large good quality fruit. • 'Krupnopladnaya' is supposedly the largest of the species. • 'Ananasnaya' is a cultivar from Russia.

If you need assistance identifying cultivars suitable for your area, you may want to contact your local Cooperative Extension office. They have experts who may provide you with information you are seeking.

HOW TO GROW A KIWIFRUIT PLANT. Unless you purchased a self-fertile kiwifruit cultivar, you will need to plant at least one male and female plant if you want your female plant to bear fruit. A general rule is to plant one male for every five female plants. Also bear in mind that mature plants spread 10 to 15 feet so you need to consider how much space you have to grow the plants. Select a location with full sun since it is required for fruit production. The plants should be planted in deep, fertile, well draining soil. Well draining soil is very important because the plants are subject to root rot from *Phytophthora*.

The best time to plant kiwifruit plants is in the spring after all danger of frost has passed. The plants should be planted at minimum 10 feet apart. Dig a hole slightly larger than the container the plant came in and about as deep. Remove the plant from the container, loosening or straightening out any overcrowded roots. The plants should be planted just deep enough to cover the roots well with soil. Backfill the hole with the original soil. Water thoroughly.   The plants should be watered regularly until they become established, but not over watered. And they need to be regularly watered during the summer.

Like grapes, kiwifruit plants need to be regularly pruned, trained, and trellised. The vines need to be pruned during the winter, and during the growing season. Proper pruning is critical for successfully growing kiwifruit. For information on properly pruning and trellising the vines, you should read the following documents *Growing Kiwifruit* from the Pacific Northwest Extension Service (https://catalog.extension.oregonstate.edu/sites/catalog/files/project/pdf/pnw507.pdf), *Kiwifruit Production Guide* from the Alabama Cooperative Extension System (http://www.aces.edu/pubs/docs/A/ANR-1084/ANR-1084.pdf) and *Kiwi Production in Oklahoma* from the Oklahoma Cooperative Extension Service (http://pods.dasnr.okstate.edu/docushare/dsweb/Get/Document-1038/F-6249web.pdf).

How to Grow from Seed. As mentioned earlier, growing kiwifruit plants from seed is generally not recommended since you will not know the sex of the plant that will sprout, and because the plant may not bear true to type. However, if you love the challenge of seeing if you can grow a kiwi plant from seed, there are some slightly different methods commonly recommended. The first thing you need to do is separate the seeds from the kiwifruit pulp. This can be done by placing the kiwi seeds with water into a blender, then straining out the seeds through a colander or fine mesh screen. The next step differs: Some recommend the seeds be placed on a moist paper towel placed inside a clear plastic bag and placed in a warm location until the plants sprout. Others recommend soaking the seeds in a container filled with water (changing the water daily) and placing it in a warm location until the seeds begin to open before placing them on a moist paper towel inside a clear plastic bag. Once the seeds have sprouted, tear pieces of the paper towel apart and plant them in moist soil in pots.

Horticulturalist Julia Morton writes the seeds should be mixed with moist sand, placed in an airtight container and placed in the refrigerator below freezing temperatures for two weeks. Then the seed/sand mixture can be planted directly into the ground, or into nursery flats. Seeds will germinate in two to three weeks.

▶ USDA Hardiness Zones: 3–9 depending upon cultivar. Chilling Hours: 50–800 hours depending upon the cultivar. Water Requirements: Significant.

How to Consume. Kiwifruit are most often eaten fresh. Ripe kiwifruit are soft to the touch. To eat fresh, you can use a knife to slice the fruit in half, then scoop out the pulp with a spoon. Or you can simply peel away the fuzzy skin. Fresh kiwi can be used in fruit smoothies, fruit salads, gelatins, etc. You'll also find recipes that use fresh kiwi in desserts, such as kiwi lime pie. Kiwifruit can be made into a jam, or mixed with other fruit like strawberry and made into a jam. Be sure to use only safe recipes from reputable sources like pectin companies, canning companies, and the National Center for Home Food Preservation.

Kiwifruit can also be dehydrated. To dry kiwifruit, peel away the skin. Slice the fruit into three-eighths to one-half inch slices. Place in an electric dehydrator and dry at 140 degrees Fahrenheit. Average drying time is five to 12 hours. To test for dryness, cut a couple of pieces in half. You should not be able to squeeze any moisture out of them, and there should be no visible moisture. And if you fold a piece over, it should not stick to itself. Once the fruit is dry, let it cool for an hour. Then package the fruit, preferably in a freezer container or glass jar. Freezer bags can also be used but unfortunately they are not rodent proof. Store the container in a dark, dry, cool place for no more than a year.

ADDITIONAL INFORMATION. The popular fuzzy skinned green pulped kiwifruit was originally classified as *Actinidia chinensis* but was later changed to *Actinidia deliciosa*. However, *Actinidia chinensis* was kept for another species name in the genus. Definitely confusing!

REFERENCES

"*Actinidia deliciosa.*" Missouri Botanical Garden. www.missouribotanicalgarden.org/Plant Finder/PlantFinderDetails.aspx?taxonid=275437&isprofile=0. Accessed 10/07/18.

Bovshow, Shirley. "How to Grow Kiwi from Store Bought Kiwi Fruit!" Foodie Gardener. http://foodiegardener.com/how-to-grow-kiwi-from-store-bought-kiwi-fruit/. Accessed 10/07/18.

California Kiwifruit website. www.kiwifruit.org. Accessed 10/07/18.

Ferguson, A.R. New temperate fruits: *Actinidia chinensis* and *Actinidia deliciosa.*" p. 342–347. In: J. Janick (ed.), Perspectives on new crops and new uses. Alexandria, VA: ASHS Press, 1999. www.hort.purdue.edu/newcrop/proceedings1999/v4-342.html#cultivars. Accessed 10/12/18.

"Growing Hardy Kiwi Vine in the Garden." Gardening Know How. www.gardeningknowhow. com/edible/fruits/kiwi/growing-hardy-kiwi-vine.htm. Accessed 10/07/18.

"Hardy Kiwi in the Home Fruit Planting." August 2016. Pennsylvania State University Extension. https://extension.psu.edu/hardy-kiwi-in-the-home-fruit-planting. Accessed 10/12/18.

"How to Grow a Kiwi Plant from Seed." iCreative Ideas. www.icreativeideas.com/how-to-grow-a-kiwi-plant-from-seed/. Accessed 10/07/18.

Ingels, Chuck A., Pamela M. Geisel, and Maxwell V. Norton, technical editors. *The Home Orchard: Growing Your Own Deciduous Fruit and Nut Trees.* Publication 3485. Richmond, CA.: University of California, Agriculture and Natural Resources Communication Services, 2007.

"Kiwifruit." University of Wisconsin, LaCrosse. http://bioweb.uwlax.edu/bio203/s2012/cejka_laur/habitat.htm. Accessed 10/07/18.

"Kiwifruit (*Actinidia deliciosa*)." The Backyard Orchard, University of California. http://homeorchard.ucanr.edu/Fruits_&_Nuts/Kiwifruit/. Accessed 10/07/18.

"Kiwifruit *Actinidia deliciosa.*" California Rare Fruit Growers. www.crfg.org/pubs/ff/kiwifruit.html. Accessed 10/07/18.

"Kiwifruit cultivars." February 16, 2007. Northwest Berry & Grape Information Network, Oregon State University, University of Idaho, Washington State University, and USDA Agriculture Research Service. http://berrygrape.org/kiwifruit-cultivars/. Accessed 10/12/18.

"Kiwifruit in California." University of California, Davis, Fruit and Nut Research and Information Center. http://fruitandnuteducation.ucdavis.edu/fruitnutproduction/Kiwi/. Accessed 10/07/18.

"Kiwifruit Production Guide." ANR-1084. June 1998. Alabama Cooperative Extension System. www.aces.edu/pubs/docs/A/ANR-1084/ANR-1084.pdf. Accessed 10/07/18.

Kolstrom, Fred. "Fruitless Kiwi." *Organic Gardening*, vol. 37, no. 9, December 1990, p. 22.

Lui, Kevin. "This Kiwifruit Isn't from New Zealand at All. It's Chinese, and This Is How It Got Hijacked." *Time Magazine*, February 9, 2017. http://time.com/4662293/kiwifruit-chinese-gooseberry-new-zealand-history-fruit/. Accessed 10/07/18.

Mattern, Vicki. "Hardy kiwifruits." *Organic Gardening*, vol. 40, no. 1, January 1993, p. 60–65.

Morton, Julia F. *Fruits of Warm Climates*. Eugene, OR: Wipf and Stock Publishers, 2003.

Nesco Dehydrator & Jerky Maker Recipes and Instructions. Two Rivers, WI: The Metal Ware Corporation, 1998.

Ortho All About Citrus and Subtropical Fruits. Des Moines, IA: Meredith Books, 2008.

"Plant of the Week: Kiwi." University of Arkansas Cooperative Extension Service. www.uaex.edu/yard-garden/resource-library/plant-week/kiwi-11-9-07.aspx. Accessed 10/09/18.

Saliyan, Tripthi, et al. "A Review of Actinidia deliciosa." *International Journal of Pharma and Chemical Research*, vol. 3, no. 1, January—March 2017, p. 103–108.

Strik, Bernadine, and Helen Cahn. *Growing Kiwifruit*. PNW 507. Pacific Northwest Extension Service. 1998. https://catalog.extension.oregonstate.edu/sites/catalog/files/project/pdf/pnw507.pdf. Accessed 10/06/18.

"Types of Kiwi for Zone 3: Choosing Kiwi for Cold Climates." Gardening Know How. www.gardeningknowhow.com/garden-how-to/gardening-by-zone/zone-3/zone-3-kiwi-plants.htm. Accessed 10/12/18.

Vossen, Paul M., and Deborah Silver. "Growing Temperate Tree Fruit and Nut Crops in the Home Garden and Landscape." University of California Cooperative Extension. http://homeorchard.ucdavis.edu/plant_Kiwi.pdf. Accessed 10/8/18.

Wall, Clint, et al. "Vegetative and Floral Chilling Requirements of Four New Kiwi Cultivars of *Actinidia chinensis* and *A. deliciosa*." *HortScience*, vol. 43, no. 2, June 2008, p. 644–647.

Walliser, Jessica. "Growing Kiwifruit: It's Easier Than You Think." *Savvy Gardening*. https://savvygardening.com/growing-kiwi-fruit/. Accessed 10/07/18.

Westover, Jessica. "How to Plant a Kiwi Actinidia." SF Gate. https://homeguides.sfgate.com/plant-kiwi-actinidia-64693.html. Accessed 10/07/18.

Wetherbee, Kris. "Kiwi and Gooseberries." *Mother Earth News*, December/January 2003, p. 83.

Whitworth, Julia. "Kiwifruit Production in Oklahoma." HLA 6249. Oklahoma Cooperative Extension Service. http://pods.dasnr.okstate.edu/docushare/dsweb/Get/Document-1038/F-6249web.pdf. Accessed 10/07/18.

Kumquat (*Fortunella*)

Kumquats, or cumquats, are small oblong or round (depending upon the species) orange colored fruit about the size of large olives. Measuring from five-eighths to 1½ inches wide, it bears a very thick yellow to deep orange peel. There's very little pulp. The peel is sweet and the pulp is tart. The entire fruit is edible. Its taste is described as a unique mix of sweet and tart incomparable to other fruit. (Personally, the Nagami kumquat tasted exactly like I expected it to, like a thin sweet orange peel with a barely noticeable center acidic pulp.)

KUMQUAT LORE. In Asian culture, citrus fruits like kumquats are supposed to bring good luck and prosperity and are frequently gifted during the New Year. In some Asian countries, kumquats are used in folk medicine to cure colds, coughs, and respiratory inflammation. According to a story on the Popsugar website, Oprah Winfrey became friends with Doria Ragland, the

Kumquat *(Tim Reckmann/Wikimedia Commons)*

mother of Meghan, the Duchess of Sussex, and gifted her with a bag of kumquats from her tree, because Ms. Ragland supposedly loves them.

TREE HISTORY. Originating in China, kumquats were described in Chinese literature as early as 118 BCE. They appeared on a list of plants grown in Japan in 1712. It's believed Chinese immigrants brought the kumquat to Hawaii, perhaps as early as 1825. They were introduced into Europe in 1846. The kumquat was introduced into California about 1880 and into Florida from Japan in 1885.

TREE CHARACTERISTICS. Originally classified in the citrus botanic genus as *Citrus japonica*, in 1915 they were reclassified into their own genus, *Fortunella*, named after Scottish botanist Robert Fortune, who introduced the kumquat into Europe. Scientist Robert Swingle reclassified kumquats into their own genus because of structural differences in their leaves, blossoms and fruit from other citrus. *Fortunella* is now the generally accepted genus for the kumquat. Instead of cultivars, the various kumquats are classified as individual species. The most popular species grown in the United States include the *Meiwa* (*Fortunella crassifolia*), *Nagami* (*Fortunella margarita*), *Marumi* (*Fortunella japonica*) and the *Hong Kong* (*Fortunella hindsii*).

Kumquat trees are fairly small evergreen, subtropical trees only growing eight to 15 feet tall and five to eight feet wide. They are slow growing trees with dense branches that sometimes have small thorns. Its dark green glossy leaves usually measure 1¼ to 3⅓ inches long. The blossoms are white and similar to citrus blossoms but less fragrant. The tree is self-fertile. Mature trees can produce hundreds of fruit annually. The trees have a longevity of 50 to 150 years.

Trees are seldom propagated by seed because the roots of trees grown from seed are generally considered weak and the tree inferior. Tree propagation is commonly done by bud grafting on trifoliate orange (*Poncirus trifoliata*) rootstock, and grapefruit and rangpur lime rootstock. Kumquat trees can also be propagated by root cuttings and air layering. Trees vegetatively propagated usually begin bearing fruit within three years. Trees grown from seed generally take seven or more years before beginning to bear fruit. The trees are often grown as hedges, ornamental trees, in containers, and even as bonsai trees.

▶ FAMILY: Rutaceae. GENUS: *Fortunella*. SPECIES: (many different species).

SELECTING A TREE. First, choose the species of kumquat tree you desire. You have a number of options depending upon your taste. Your options include, but are not limited to the following: '*Meiwa*'—produces a round fruit with a sweet rind and flesh. • '*Marumi*'—produces round fruit with sweet peel and tart pulp. • '*Nagami*'—produces oval fruit with a thinner sweet rind and tart pulp, little juice. • '*Changshou*' *(Fukushu)*—pear shaped fruit with sweet rind and mildly tart juicy pulp. • '*Hong Kong*'—thin skinned, very tart, large seeds, makes a better ornamental or bonsai tree than edible fruit tree.

When selecting a tree from the nursery, check the tree's overall health. Look for active growth, such as new or young leaves. Make sure its leaves are not yellow or discolored, and free from insects or signs of insect damage. Check the entire tree to make sure it is free from injury, such as broken branches, cuts in the trunk, etc. Check to make sure the tree trunk is not sunburned. And if possible, gently lift the tree out of the pot to check the roots. Overgrown trees that have seriously girdled roots will have a poor root structure so they should be avoided.

HOW TO GROW A KUMQUAT TREE. Kumquats are most productive in full sun, but can tolerate some shade. The trees also tolerate a wide range of soils, provided they are well draining. To plant a tree, dig a hole at least twice the diameter of the pot the tree came in, about as deep as the container. Place the tree in the hole, loosening or cutting away any girdled roots. The tree should be planted so that the top of the root ball is about one to two inches higher than the soil level. Refill the hole with the original soil. Water the newly planted tree thoroughly. The tree should be regularly watered the first two years, but not over watered.

Young citrus trees are sensitive to frost. If temperatures are expected to dip to 29 degrees Fahrenheit or lower, you need to protect the tree from frost. You should wrap the tree trunk and branches in insulation material like cardboard, fiberglass or frost protection covers. Make sure the soil is moist and not dry, since damp soil retains and radiates more heat than dry soil. Bare soil also radiates more heat than soil covered with mulch or other ground covers. If your tree does sustain frost damage, don't prune away any dead branches until the spring. This allows time for the tree to recover in warmer weather. It also allows you to better identify the damaged branches to be removed.

Kumquat trees benefit from fertilization to keep them healthy and producing. Apply an organic fertilizer or a citrus fertilizer as directed on the package. Pruning is only required to remove dead branches and if you want to keep the tree in specific height. The main pest of kumquat trees is mealy bugs. They are also a host for fruit flies. Harvesting kumquats is not the same as harvesting oranges. You can't just pull them off the tree because doing so leaves an open scar on the fruit, leading it to drying out and spoiling quickly. Instead, they need to be clipped from the tree.

▶ USDA HARDINESS ZONES: 9–10. CHILLING HOURS: 0. WATER REQUIREMENTS: Average.

How to Consume. Ripe kumquats can be eaten fresh whole. To select a ripe kumquat, look for ones that are bright orange or yellow orange and firm to touch. Wash it thoroughly. Eat it whole, or depending upon the species, you may want to eat just the peel if the pulp is particularly sour with lots of seeds.

Fresh kumquats can be stored in the refrigerator (35 to 40 degrees Fahrenheit) for up to three weeks in an airtight container, and four to six months in the freezer (zero degrees Fahrenheit). Fresh kumquats can also be sliced and added to salads, or used to season cooked meats and fish. Fresh kumquats are often infused into vodka. To infuse vodka with kumquats, first wash and slice a number of kumquats into halves or quarters. Place the fruit into an airtight jar, like a canning jar. Pour the vodka over the fruit. Seal the jar and shake it a few times. On a daily basis shake the jar. Let the fruit infuse, i.e., let the jar sit for five days to two weeks, depending upon how strong you want the flavor to be. When it's ready, strain the fruit from the alcohol and enjoy. Kumquats are frequently candied, and made into marmalade.

Additional Information. The name kumquat stems from Cantonese words meaning "golden orange." In the industry, the kumquat has been nicknamed the "little gold gem of the citrus industry." Kumquats are high in vitamin

C. Studies conducted on mice indicate kumquats may help reduce obesity, lower blood sugar, and boost the immunity system. Many people first heard the word kumquat when W.C. Fields said "How about a kumquat, my little chickadee?" in the 1940 film *My Little Chickadee*.

REFERENCES

Andreu, Michael G., Melissa H. Friedman, and Robert J. Northrop. *Fortunella spp., Kumquat*. University of Florida, Institute for Food and Agricultural Sciences Extension. 2012.

Citrus for Hawai'i's Yards and Gardens. F&N-14. College of Tropical Agriculture and Human Resources, University of Hawai'i at Manoa. June 2008. www.ctahr.hawaii.edu/oc/freepubs/pdf/F_N-14.pdf. Accessed 07/20/18.

"*Fortunella* Swingle." USDA, NRCS. 2018. The PLANTS Database. National Plant Data Team, Greensboro, NC 27401-4901. https://plants.sc.egov.usda.gov/core/profile?symbol=FORTU. Accessed 08/06/18.

"Fruit of the Month: Kumquat." Food Forward, Inc. March 26, 2013. https://foodforward.org/2013/03/fruit-of-the-month-kumquat/. Accessed 08/07/18.

Geisel, Pamela M., and Carolyn L. Unruh. *Frost Protection for Citrus and Other Subtropicals*. ANR Publication 8100. University of California, Division of Agriculture and Natural Resources, 2003.

Golden Gems Kumquat. Florida Department of Agriculture and Consumer Services. N.d. www.freshfromflorida.com/content/download/61407/1283755/Kumquat.pdf. Accessed 08/06/18.

Karp, David. "Kumquats: Sweet-tarts of the citrus rodeo." *Los Angeles Times*. California Living Section. February 25, 2009. www.latimes.com/style/la-fo-kumquat25-2009feb25-story.html. Accessed 08/07/18.

"Kumquat." Wisconsin Department of Public Instruction. n.d. https://dpi.wi.gov/sites/default/files/imce/school-nutrition/pdf/fact-sheet-kumquat.pdf. Accessed 08/06/18.

"Kumquat (*Citrus japonica*)." FGCU Food Forest Plant Database. Fort Myers, FL: Florida Gulf Coast University. 2015. https://www2.fgcu.edu/UndergraduateStudies/files/Kumquat.pdf. Accessed 08/06/18.

"Kumquats." Houston, TX: Urban Harvest. n.d. http://urbanharvest.org/documents/118591/5124168/Kumquats+2016+-+Edited.pdf/9d64d87e-5051-41be-9738-c982fda36750. Accessed 08/06/18.

"Kumquats Are Tiny Citrus to Use in Relish or Marmalade." University of California, Agriculture and Natural Resources. n.d. http://ucanr.edu/sites/fresnonutrition/files/14352.pdf. Accessed 08/09/18.

Love, Ken, Richard Bowen, and Kent Fleming. *Twelve Fruits with Potential Value-Added and Culinary Uses*. Manoa, HI: College of Tropical Agriculture and Human Resources, University of Hawai'i at Manoa. 2007. www.ctahr.hawaii.edu/oc/freepubs/pdf/12fruits.pdf. Accessed 08/09/18.

McCulloch, Marsha. "What Are Kumquats Good for and How Do You Eat Them?" *Healthline*. July 2, 2018. www.healthline.com/nutrition/kumquat. Accessed 08/10/18.

Morton, Julia F. *Fruits of Warm Climates*. Eugene, OR: Wipf and Stock Publishers, 2003.

"Oprah Winfrey Quotes About Royal Wedding." PopSugar. www.popsugar.com/celebrity/Oprah-Winfrey-Quotes-About-Royal-Wedding-June-2018-44936354. Accessed 08/10/18.

Ortho All About Citrus and Subtropical Fruits. Des Moines, IA: Meredith Books, 2008.

Penner, Karen P. "Refrigerator/Freezer Approximate Storage Times." L-805. Cooperative Extension Services, Kansas State University, Manhattan. March 1990. https://nchfp.uga.edu/how/store/ksu_refrig_freeze.pdf. Accessed 08/07/18.

Piper, Jacqueline M. *Fruits of South-East Asia: Facts and Fiction*. Oxford, NY: Oxford University Press, 1989.

Sauls, Julian W. *Home Fruit Production—Miscellaneous Citrus*. Cooperative Extension, Texas A&M University. December 1998. https://aggie-horticulture.tamu.edu/citrus/miscellaneous.htm. Accessed 08/07/18.

"SelecTree: Tree Detail KUMQUAT." California Polytechnic University. https://selectree.
    calpoly.edu/tree-detail/citrus-japonica. Accessed 08/06/18.
Swain, Roger B. "Kumquats." *Horticulture*, vol. 96, no. 1, January/February 1999, p. 71–74.

Langsat (*Lansium domesticum* Corr.)

Also known as lanzones, langsat is a slightly oval shaped fruit usually measuring about one to two and a half inches in diameter. It has a somewhat soft
slightly hairy green thin skin that turns yellowish when ripe. The skin possesses a whitish sticky latex when split or broken. When the skin is peeled
away, you'll find a somewhat white translucent juicy pulp divided normally
into five segments, give or take one. You'll find one to five small hard inedible
green seeds in each fruit. The fruit is slightly acidic. Hence its taste is often
described as a mild sweet grapefruit.

LANGSAT LORE. There's a popular myth in the Philippines about how the
langsat was converted from a poisonous fruit to an edible one. There are different versions of the story, but basically it involves a thief who, being pursued
by residents, hid in an area with lots of langsat trees. Unable to safely sneak
out of the area, he became hungry and picked and ate a langsat. Upon eating
the fruit he immediately died, and the people believed the langsat was poisonous.
    Years later a woman visits the area. And this is where the myth differs
somewhat. In one version, the woman is a beautiful woman. She strolls to a
langsat tree, picks a fruit, pinches it then eats it. To everyone's amazement
she doesn't die. She tells the villagers when she pinched the fruit, she removed
the poison and that langsat fruit were now safe to eat. The villagers believed
she had magical powers or was a fairy because she made all langsat fruit edible.
    In another version, the woman is a poor old woman taken in by the vil-
lagers. Upon seeing the langsats growing on the trees, she tells the villagers
they are edible. To prove it, she eats a few and to everyone's amazement, doesn't
die. She told them that if you pinched the fruit before eating it, the white
sticky sap that was released counteracted the poison making it edible. She
stressed how important it was to properly eat the fruit.
    According to lore, langsat, a.k.a. lazones, got its name from the Filipino
word *lason*, which means poison. According to folk medicine, langsat is
believed to prevent cancer. Although the fruit has antioxidants, there are no
scientific studies to substantiate this lore. The fruit peel supposedly can help
with diarrhea and intestinal parasites. And the seeds, when ground into a
powder is said to reduce a fever. But there is also a lack of scientific studies
to support these claims.

Langsat *(Izham Alias/Wikimedia Commons)*

TREE HISTORY. The langsat tree originated in Southeast Asia in the area from Peninsular Thailand in the west, to Borneo in the east. It was introduced into the Philippines during prehistoric times. Trees can also be found in Vietnam, Surinam, Sri Lanka, Burma, and India. In 1927 trees were introduced into Honduras. Sometime before 1930 the langsat was introduced into Hawaii and St. Croix. It was also introduced into Puerto Rico. Trees are also growing in Australia.

TREE CHARACTERISTICS. Considered an ultra tropical tree, trees grown in the wild can grow up to 90 feet tall. However cultivated trees usually only grow up to 30 feet tall. The trees usually have a straight trunk, with grayish brown furrowed bark measuring up to two and a half feet in diameter. The trunk also contains a sticky sap.

Langsat trees have upright branches. The trees feature dark green leaves 9 to 20 inches long, with alternate leaflets. The leaflets measure 6 to 8 inches long and two and a third to four inches wide. The leaves are shiny on the top and slightly hairy on the underside. The tree bears pale yellow blossoms, not quite half an inch wide. The blossoms have a sweet scent. The blossom clusters grow from large leafless branches, and the tree trunk. The blossoms are borne in clusters of 20 to 30 on a stem. The tree is self pollinating. When the fruit emerge, they appear in grape like clusters. The trees have shallow roots. They have lots of surface-feeding roots.

Trees are propagated by seed, by grafting, and to a lesser extent by air layering, budding and softwood cuttings. Langsat trees grow slowly. Trees started from seed generally take 10 to 15 years to start fruiting. While grafted trees usually take eight years to start bearing fruit.

▶ FAMILY: Meliaceae. GENUS: *Lansium* Correa. SPECIES: *Lansium domesticum* Correa.

SELECTING A TREE. If there's a specific variety you enjoy consuming, then you may want to consider selecting a vegetatively propagated tree because they are truer to type than those trees grown from seed. Another consideration in selecting a tree is your patience level in waiting for the tree to begin bearing fruit. Remember that grafted trees usually begin bearing fruit sooner than trees grown from seed. When selecting a tree from the nursery, check the tree's overall health. Make sure its leaves are not discolored, and that they are free from insects or signs of insect damage. Check the entire tree to make sure it is free from injury, such as broken branches, cuts in the trunk, etc. If possible, gently lift the tree out of the pot to check the roots. Overgrown trees that have seriously girdled roots will likely develop a poor root structure so they should be avoided.

HOW TO GROW A LANGSAT TREE. Langsat trees grow best in moist, rich, well drained sandy loam to clay loam soils with a pH of 5.3 to 6.5. It does not grow well in alkaline soils. It also does not withstand cold temperatures. At temperatures below 40 degrees Fahrenheit the tree will defoliate. When planting a tree, make sure to dig a hole at least twice as wide as the tree's root ball, and at least as deep as the root ball. Carefully place the tree in the hole, unfurling any girdled roots. Refill the hole with 50 percent of the original soil and a 50 percent mix of organic matter and fertilizer, like compost and manure. Water the tree thoroughly.

During the first two to three years, the tree should be provided with 50 percent shade to prevent sun scalding. Be sure to water the tree regularly especially during dry periods. During the first year the tree should also be fertilized with a complete inorganic fertilizer (one pound total) and an organic fertilizer (ten pounds) split into two to four applications. A major pest of langsat trees is a bark borer. To control the pest, infested bark should be immediately removed and destroyed. Expose any borer tunnels to sunlight. Other pests include scales, mealybugs, beetles and moths. The fruit is also a host for fruit flies.

How to Grow from Seed. Retrieve only the large seeds from the langsat since they have higher vigor than small seeds. Clean away any pulp from the seed and soak it in water for one to two days to soften the seed shell. Since seeds only remain viable for seven to ten days, and even less if they dry out, plant them as soon as possible. Plant the seed half-inch deep in moist soil. The seeds will germinate in two to three weeks.

▶ USDA Hardiness Zone: 11. Chilling Hours: 0. Water Requirements: 40 to 180 inches rainfall annually.

How to Consume. Langsat are usually eaten fresh. Either peel away the skin, or squeeze the fruit until the skin splits and any seeds are loosened from the fruit segments. Enjoy the pulp. Ripe langsat pulp is sometimes preserved in syrup.

Additional Information. On the island of Java dried langsat peels are burned to repel mosquitoes. In some countries, the bark has been used in folk medicine. It's ground into a powder and used in poultices to treat scorpion stings. The tree wood has been used to make furniture.

References

Bareja, Ben. "Crop Info and How-To Guide in Growing Lanzones." Cropsreview.com. 2010. www.cropsreview.com/lanzones.html. Accessed 06/07/18.

City Agriculturalist Office, Davao, Philippines. "Lanzones Production." 2010. http://agriculture. davaocity.gov.ph/wp-content/.../lanzonesproduction20120217140416.pdf. Accessed 06/11/18.

Damsker, Matt. "The Wild Manga of Borneo!" *Organic Gardening*, vol. 39, no. 3, March 1992, p. 20.

Ecosystems Research and Development Bureau. "Pesticidal/Insecticidal Plants." Research Information Series on Ecosystems, vol. 27, no. 2, May—August 2015. Ecosystems Research and Development Bureau, Department of Environment and Natural Resources, College 4031, Laguna.

Fern, Ken. "Lansium domesticum." Tropical Plants Database. http://tropical.theferns.info/view tropical.php?id=Lansium+domesticum. Accessed 06/08/18.

"Langsat." World-Crops.com. http://world-crops.com/langsat/. Accessed 06/07/18.

"Lansium domesticum." EcoCrop. Food and Agriculture Organization of the United Nations. http://ecocrop.fao.org/ecocrop/srv/en/cropView?id=2318. Accessed 06/29/18.

"*Lansium domesticum* Corr." *Agricultural ID Aid Manual*. California Department of Food and Agriculture. n.d. www.cdfa.ca.gov/plant/pe/AgCommID/page25.htm. Accessed 06/08/18.

"*Lansium domesticum* Correa." USDA, NRCS. 2018. The PLANTS Database. National Plant Data Team, Greensboro, NC 27401-4901. https://plants.usda.gov/core/profile?symbol= LADO2. Accessed 06/07/18.

"*Lansium domesticum* (Meliaceae)." Montoso Gardens. www.montosogardens.com/lansium_ domesticum.htm. Accessed 06/08/18.

Loquias, Virgilio L. "Production Guide for Lanzones." Philippine Department of Agriculture, Bureau of Plant Industry, Manila, Philippines. n.d. http://bpi.da.gov.ph/bpi/images/ Production_guide/pdf/LANZONES.pdf. Accessed 06/12/18.

Morton, Julia F. *Fruits of Warm Climates*. Eugene, OR: Wipf and Stock Publishers, 2003.

Orwa, C., et al. "*Lansium domesticum* Correa Meliaceae." Agroforestry Database: A Tree Reference and Selection Guide version 4.0. www.worldagroforestry.org/treedb2/AFTPDFS/Lansium_domesticum.PDF. Accessed 6/07/18.

Paull, Robert E. *Longkong, Duku, and Langsat: Postharvest Quality-Maintenance Guidelines.* F_N-42. College of Tropical Agriculture and Human Resources, University of Hawaii at Manoa. June 2014. www.ctahr.hawaii.edu/oc/freepubs/pdf/F_N-42.pdf. Accessed 06/08/18.

Piper, Jacqueline M. *Fruits of South-East Asia: Facts and Fiction.* Oxford, NY: Oxford University Press, 1989.

Popenoe, Wilson. *Manual of Tropical and Subtropical Fruits: Excluding the Banana, Coconut, Pineapple, Citrus Fruits, Olive and Fig.* NY: Macmillan Co., 1920.

Tillar, Martha, et al. "Revisión de *Lansium domesticum* Corrêa y sus usos en cosmética." *Boletín Latinoamericano y del Caribe de Plantas Medicinales y Aromáticas*, vol. 7, n. 4, 208, p. 183–189.

Watson, Brian. "Langsat and Duku." *The Archives of the Rare Fruit Council of Australia.* May 1982.

Whitman, William F. "Growing and Fruiting the Langsat in Florida." *Proceedings of the Florida Horticulture Society*, vol. 93, 1980, p. 136–140.

"WRA Species Report: Lansium domesticum." U.S. Forest Service, Pacific Island Ecosystems at Risk (PIER). www.hear.org/pier/wra/pacific/Lansium%20domesticum.pdf. Accessed 06/08/18.

Longan (*Dimocarpus longan* Lour.)

Also known as a lungan and a dragon's eye, longans are closely related to lychees and are in the same botanic family (*Sapindaceae*). Round fruits measuring about half to one inch in diameter, they are slightly smaller than lychees. They bear a thin somewhat rough leather-textured outer skin that starts out green but turns tan or yellowish brown when ripe. Like the lychee, the pulp is a translucent white. Its taste is similar to that of the lychee, having a grape-like flavor, except not as sweet a lychee. Each longan has a large shiny dark brown or black seed, depending on the variety, in the middle of the white pulp, giving it the appearance of an eye. Because of its appearance, the fruit is also known as a dragon's eye. The seed is easily removed from the fruit.

LONGAN LORE. In Vietnam there's a myth that if you press the dragon's eye seed against a snake bite, it will draw out and absorb the venom. And one can guess, this has proven not to work. Fresh longans are believed to reduce fevers. They are also said to cure a variety of conditions ranging from amnesia to stomach problems. But there is currently a lack of scientific studies to support these claims.

TREE HISTORY. The longan tree is native to Southern China. Although a few early horticulturalists listed its origin as India, most experts today believe it is indeed native to China. The first written record of the tree dates it back to at least 200 BCE when it appeared in the Han Dynasty (206–220 BCE). It was introduced into India in 1798, into Australia in the mid–1800s, and Thailand

Longan *(Dezidor/Wikimedia Commons)*

in the late 1800s. The U.S. Department of Agriculture introduced it into Florida in 1903.

TREE CHARACTERISTICS. An evergreen tree, under ideal conditions the longan tree can grow up to 50 feet tall. But trees grown in the U.S. generally grow 25 to 35 feet. The tree's canopy tends to be as wide as the tree is tall. Its crown is round or slightly oblong. The tree's trunk bears rough corky like bark. The tree's trunk and branches are brittle. The tree has dark green shiny ovate leaves measuring generally four to eight inches long, but can be up to 12 inches long. The under side of the leaves are grayish green and have slight minuscule hairs. The leaves are arranged like the pattern of a feather, having from four to ten leaflets arranged on each side of a single stalk. Newly emerging leaves are a dark shade of red. The tree bears small pale yellow blossom clusters. The tree contains both male and female blossoms. Although some self pollination occurs, pollination is generally done by insects. Studies have shown that honey bee pollination increases the fruit yield.

Longan trees are primarily propagated vegetatively, by air layering, grafting, and at times from cuttings. Trees propagated by seed are usually just used as rootstock since they do not bear fruit true to type. Trees propagated from seed can take up to seven years before bearing fruit. Whereas trees propagated by air layering usually begin bearing fruit in three to four years.

There are many longan cultivars, but popular cultivars vary by country. In Florida, where most of the longans in the United States are grown, more than 90 percent of the trees grown are the 'Kohala' cultivar. The 'Kohala' is also widely grown in California, and in Hawaii, along with the 'Egami.' 'Kohala' is also the most popular cultivar in Australia. In Thailand popular cultivars include the 'Daw,' 'Chompoo,' 'Biew Khiew,' and 'Dang.' In China, the popular cultivars vary by region, but include the 'Dawuyuan,' 'Chuliang,' 'Fuyan,' and 'Wulongling.'

▶ FAMILY: Sapindaceae. GENUS: *Dimocarpus* Lour. SPECIES: *Dimocarpus longan* Lour.

SELECTING A TREE. When selecting a tree, you should consider those cultivars that are commonly grown in your region since they have demonstrated their hardiness and productivity in the area. In the United States, consider the 'Kohala' cultivar which is popular in Hawaii, California and Florida. When selecting a tree from the nursery, check the tree's overall health. Make sure its leaves are not discolored, and free from insects or signs of insect damage. Check the entire tree to make sure it is free from injury, such as broken branches, cuts in the trunk, etc. Check to make sure the tree trunk is not sunburned. Also visually inspect the pot soil to make sure there is no fungi growing from the soil, indicating uneven or over watering that can affect the overall health of the trees. If possible, lift the tree out of its pot to check to make sure its roots are not seriously overgrown and girdled, since they are likely to develop a poor root system.

HOW TO GROW A LONGAN TREE. Longans grow best in rich sandy-loam soils, but tolerate a wide range of soils. They prefer a pH of 5.0 to 6.5. They do not tolerate temperature below 32 degrees Fahrenheit. Young trees suffer leaf and twig damage at 30 to 31 degrees Fahrenheit. And temperatures of 28 degrees Fahrenheit and lower can kill or cause major damage to young trees. To plant a tree, select a sunny location that can accommodate the size of a mature longan tree. Dig a hole at least twice the diameter of the pot the tree came in, and at least as deep as the container. Place the tree in the hole, filling the hole with the original soil. The tree should be planted so that the root ball is about one to two inches higher than the soil level/grade. Water thoroughly.

Because of the weak stem of young seedling trees, and because trees propagated by air layering don't have tap roots, young longan trees should be staked their first year or two to prevent any potential wind damage. To properly stake a tree, place two stakes in the ground. They should be placed on opposite sides of the tree, making sure they are outside of the root ball. Next, determine the proper height for the support ties. To do this, use your fingers and move it up the tree until the tree stands upright. Note the location, since the ties should be placed six inches above that point. If you have old pantyhose around, they make good ties since they provide the support needed, but their flexibility also allows for trunk growth.

The stakes and ties should be removed when the tap roots have grown enough to support the tree, or the tree is sturdy enough to withstand strong winds. Usually the stakes can be removed after a year or two. Make sure to remove the stakes when they are no longer needed, since leaving them in place can ultimately harm the tree with the ties and stakes causing tree injury, interfering with branch development, and stifling normal trunk growth. Minimal pruning is required. The tree needs to be pruned only to remove dead branches. You can also prune the tree if you want to keep it a specific size or shape. Potential pests include scales, aphids, mealy bugs and leaf eating caterpillars.

HOW TO GROW FROM SEED. As mentioned earlier, trees grown from seed do not bear true to type. But if you want to grow a tree from seed, be aware that longan seeds are recalcitrant. They should be planted very shortly after being removed from the fruit so they will not dry out. If the longan from which you retrieved the seed was not grown locally, but was imported from Asia, be aware that it was irradiated, thereby significantly decreasing the viability of the seed.

Upon removing the seed from the fruit, wash away any pulp. Place the seed in a pot in a moist planting medium. Place the potted seeds in a sunny location. Keep the soil moist. Germination takes seven to ten days.

► USDA HARDINESS ZONES: 8–11. CHILLING HOURS: 0. WATER REQUIRE-MENTS: Average.

HOW TO CONSUME. They are normally eaten fresh but can also be frozen or dried. Commercially they are also canned. To eat them fresh, thoroughly wash them. Use a knife to break the skin, and then peel it off. If you are in a location where a knife is not available, you can also gently bite the fruit to break the shell. You can then either squeeze out the flesh, or peel the skin away. Fresh longans can be added to fruit salads and smoothies. They are used in puddings, baked goods, and alcoholic cocktails.

Longans can be dried whole in an electric food dehydrator. Be sure to thoroughly clean them in water before placing them in the dehydrator. It can take up to 24 hours to dehydrate them. In Thailand, longans are dried by boiling the fruit for five minutes, and then oven drying them at 131 degrees Fahrenheit. Once the longans begin to dry, the temperature is raised to 158 degrees Fahrenheit until the fruit is completely dried, about 19 to 20 hours total. They can be sealed in an airtight container and stored in the refrigerator for up to a week. You should know that at temperatures less than 50 degrees Fahrenheit the skin can turn dark, but this will not affect the fruit's taste.

ADDITIONAL INFORMATION. If you are unable to find fresh longans, you can find canned longans at Asian markets. Longans are a source of vitamin C and antioxidants. Crushed seeds produce foam, which in some countries is used as shampoo. The wood is used in the construction of furniture and other articles.

REFERENCES

Campbell, B.A., and C.W. Campbell. "Preservation of Tropical Fruits by Drying," *Proceedings of the Florida State Horticultural Society*, vol. 96, 1983, p. 229–231.

Cavalli, Ellen. "Get Happy!" *Vegetarian Times*, April 2000, p. 100.

Choo, Wong Kai. *Longan Production in Asia*. RAP Publication: 2000/20. Bangkok: Food and Agriculture Organization of the United Nations, Regional Office for Asia and the Pacific, 2000. www.fao.org/docrep/003/X6908E/x6908e00.htm. Accessed 06/25/18.

Crane, Jonathan H. *Longan Growing in the Florida Home Landscape*. FC49. University of Florida, Institute of Food and Agricultural Sciences Extension. November 2016. https://edis.ifas.ufl.edu/pdffiles/MG/MG04900.pdf. Accessed 06/24/18.

Cunningham, Sally. Asian *Vegetables: A Guide to Growing Fruit, Vegetables and Spices from the Indian Subcontinent*. Bath, UK: Eco-logic Books, 2009.

"*Dimocarpus longan* Lour." USDA, NRCS. 2018. The PLANTS Database. National Plant Data Team, Greensboro, NC 27401-4901. https://plants.sc.egov.usda.gov/core/profile?symbol= DILO7. Accessed 06/26/18.

Gandy, Kathleen. "Taste of the Tropics." *Australian Gourmet Traveller*, vol. 5. no. 1, January 2005, p. 33–34.

Gomes, Andrew. "Tropical Crops Reaped $7.5M for Isle Farmers, Survey Finds." *Honolulu Star Advertiser*, September 30, 2017.

Jiang, Yueming, et al. "Postharvest Biology and Handling of Longan Fruit (*Dimocarpus longan* Lour.)" *Postharvest Biology and Technology*, vol. 26, 2002, p. 241–252.

Keogh, R., I. Mullins, and A. Robinson. *Lychee and Longan*. Pollination Aware Case Study 18. RIRDC Publication No. 10/124. Australian Government Department of Agriculture, Fisheries and Forestry, and Rural Industries Research and Development Corporation, and Horticulture Australia Limited. 2010.

Ketsa, Saichol, and Robert E. Paull. *Longan: Postharvest Quality-Maintenance Guidelines*. F_N-35. College of Tropical Agriculture and Human Resources, University of Hawaii at Manoa. 2014. www.ctahr.hawaii.edu/oc/freepubs/pdf/F_N-35.pdf. Accessed 06/18/18.

Liu, Yuge, et al. "Antioxidant Activity of Longan (*Dimocarpus longan*) Barks and Leaves." *African Journal of Biotechnology*, vol. 11, no. 27, April 3, 2012, p. 7038–7045.

"Longan." University of California Cooperative Extension, Agriculture and Natural Resources Ventura County. http://ceventura.ucanr.edu/Com_Ag/Subtropical/Minor_Subtropicals/Longan/. Accessed 06/24/18.

"Longan." Urban Harvest Inc. http://urbanharvest.org/documents/118591/436395/Longan.pdf/4790cb37-390b-4dec-8dc2-c87a4ae50e89. Accessed 06/24/18.

"Longan, also called Dragon's Eye." Rare Fruit Club (Australia). www.rarefruitclub.org.au/Level2/Longan.htm. Accessed 06/24/18.

"Longan (*Dimocarpus longan*)" FGCU Food Forest Plant Database. www2.fgcu.edu/Under graduateStudies/files/Longan.pdf. Accessed 06/24/18.

McCord, Garrett. "Oh, Little Longans." *Epicurious* [blog] March 11, 2011. www.epicurious.com/archive/blogs/editor/2011/week10. Accessed 06/25/18.

McDermott, Annette. "Longan Fruit vs. Lychee: Health Benefits, Nutrition Information, and Uses." *Healthline*. June 22, 2017. www.healthline.com/health/longan-fruit-vs-lychee-benefits. Accessed 06/24/18.

Morton, Julia F. *Fruits of Warm Climates*. Eugene, OR: Wipf and Stock Publishers, 2003.

Orwa, C., et al. "Dimocarpus longan." 2009 Agroforestree Database: a Tree Reference and Selection Guide, version 4.0. www.worldagroforestry.org/treedb/AFTPDFS/Dimocarpus_longan.PDF. Accessed 06/25/18.

Pham, V.T., M. Herrero, and J.I. Hormaza. "Fruiting Pattern in Longan, *Dimocarpus longan*: from pollination to Aril Development. *Annals of Applied Biology*, vol. 169, no. 3, November 2016, p. 357–368.

Piper, Jacqueline M. *Fruits of South-East Asia: Facts and Fiction*. Oxford, NY: Oxford University Press, 1989.

Pons, Luis. "Fruitful Studies in Puerto Rico." *Agricultural Research*, vol. 56, no. 1, January 2008, p. 12–14.

Popenoe, Wilson. *Manual of Tropical and Subtropical Fruits: Excluding the Banana, Coconut, Pineapple, Citrus Fruits, Olive and Fig*. NY: Macmillan Co., 1920.

Wall, Marisa M. "Ascorbic Acid and Mineral Composition of Longan (*Dimocarpus longan*), Lychee (*Litchi chinensis*) and Rambutan (*Nephelium lappaceum*) Cultivars Grown in Hawaii." *Journal of Food Composition and Analysis*, vol. 19, 2006, p. 655–663.

Wood, Marsha. "New Options for Lychee and Longan Fans and Farmers." *Agricultural Research*, May 2004, p. 20–22.

Yang, Xuena, et al. "Antioxidant Activities of Fractions from Longan Pericarps." *Food Science and Technology*, vol. 34, no. 2, April/June 2014, p. 341–345.

Yingling, Kimlai. "What Is the Difference Between the Lychee, Rambutan and Longan?" *Huffington Post* [blog] January 30, 2014. www.huffingtonpost.com/kimlai-yingling/lychee-rambutan-and-longan_b_4690073.html. Accessed 06/24/18.

Loquat (*Eriobotrya japonica* [Thunb.] Lindl.)

Also known as a Chinese Plum or Japanese Plum, its taste has been described as a blend of apricot, cherry and plum, and as a cross between a peach and a mango. The fruit, usually ranging from about one to two inches, can be oval, round, or somewhat pear shaped. And although flesh colors can range from white to orange, most of the loquat cultivars grown in the United States are the pale yellow or orange flesh cultivars. The most commonly grown cultivars are sweet, although tart varieties are sometimes grown. Loquats have several large seeds in the center and not much pulp in ratio to the seeds. Loquats should be harvested ripe if eaten fresh since they do not sweeten much after harvest. They have a short shelf life of less than two weeks in the refrigerator.

LOQUAT LORE. According to Chinese legend, a school of carp in a river dreamed of being able to swim up the river but lacked the strength to swim upstream. A patch of loquat trees grew near the river. When the trees started bearing fruit, some of the loquats fell into the water. The carp ate the loquats and the fruit bestowed them with such strength they were able to swim up the river to the waterfall. They had so much strength they swam all the way up the waterfall where they turned into magic dragons. Thus only Chinese royalty were allowed to eat loquats to insure commoners would not gain the power imbued in the fruit.

A lot of lore exists regarding medical uses of loquats. These uses have not been scientifically studied and thus have not been verified as medically sound. But in China loquats have been used as an expectorant. Loquat leaves are used by Chinese herbalists who claim they provide a variety of health benefits. They say a properly prepared tea made from the leaves is good for a cough, and for curing depression, diarrhea, and cancer. The leaves are also said to be beneficial in the treatment of asthma, the hiccups, and diabetes. In ancient times in Japan, loquat wood was used to make weapons. There's a myth that wounds made by a weapon made from loquat wood will fester.

TREE HISTORY. Most experts believe the loquat originated in China and was later introduced into Japan, where its cultivation has been documented back to the late 1100s. Subsequently it was introduced into France, and then England in the 1780s, and the Mediterranean in the 1800s. Chinese immigrants were believed to have brought the loquat to Hawaii, and from there, the loquat

Loquat *(Jean-Pol GRANDMONT/Wikimedia Commons)*

made its way to California in the mid 1800s, and then to Florida prior to 1887. A Bulletin of the California Agricultural Experiment Station stated by 1887 it could be found in gardens throughout the state, and by 1879 was "well distributed in the gardens of the City of Sacramento and in 1892 as being commonly grown in Butte County." In the early 1900s loquats were grown commercially in Orange County, CA. This is credited to C.P. Taft of Orange, who developed 16 different cultivars he grew in his orchard. Those cultivars included the 'Advance,' 'Blush,' 'Champagne,' 'Early Red,' 'Eulalia,' 'Golden,' 'Pineapple,' 'Premier,' 'Tanaka,' 'Thales' and others. California is the only state where small commercial loquat acreage can be found. But other states where loquats are grown include Florida, Louisiana, Georgia, Hawaii, and Puerto Rico.

TREE CHARACTERISTICS. Loquat trees can be grown for their fruit or as landscape trees. Subtropical evergreens which can grow up to 30 feet tall, most loquat trees you'll see growing in home landscapes tend to range from 15 to 25 feet. The tree crown is dense and generally about as wide as the tree is tall. Frequent branch sprouts near the trunk need to be removed if grown as a tree. If desired, they can be grown as hedges. Loquats can be grown in containers, and can also be espaliered.

The trees feature thick broad dark green, somewhat glossy leaves up to 12 inches long. The underside of the leaves is a lighter shade of green and somewhat whitish and slightly fuzzy. Tree blossoms appearing from fall to winter consist of clusters of small white flowers. The blossoms are wonderful for attracting pollinating bees to your yard during the cold weather seasons. Although loquats are fully or partially self fertile, depending upon the cultivar, cross pollination increases fruit set and size.

Sample popular cultivars include: 'Gold Nugget'—a cultivar with sweet yellow/orange pulp, self fruitful. • 'Big Jim'—sweet, light orange skin and pulp, self fruitful. • 'Thales'—a yellow pulp cultivar with a sweet/tart flavor, self fruitful. • 'MacBeth'—large fruit, yellow skinned, sweet cream colored pulp, self fruitful. • 'Champagne'—a yellow skinned, tart, juicy, white flesh cultivar, not self fruitful.

▶ FAMILY: Rosaceae. GENUS: *Eriobotrya* Lindl. SPECIES: *Eriobotrya japonica* (Thunb.) Lindl.

SELECTING A TREE. Loquats can be grown from seeds. However, if you are planning to grow a loquat tree primarily for the fruit, many recommend purchasing a loquat tree grafted on rootstock because it begins to bear fruit faster than seedling trees and is said to bear better quality fruit. Trees grown from seeds can take up to 10 years before bearing fruit. Multiple varieties of yellow and white flesh varieties exist. Some are better suited to cooler climates. When selecting a variety, check with your local nursery for varieties best suited to your area. When selecting a tree from the nursery, check the tree's overall health. Make sure its leaves are not discolored, and free from insects or signs of insect damage. Check the entire tree to make sure it is free from injury, such as broken branches, cuts in the trunk, etc. Check to make sure the tree trunk is not sunburned. If possible, lift the tree out of the pot and check its roots. Trees with significant girdled roots should be avoided.

GROWING LOQUATS. Loquat trees grow in most soils (clay, loam, sandy) provided there is adequate drainage. The shallow root systems of the trees do not tolerate excessive water. Also bear in mind when selecting a location for a loquat tree that they do best in full sun, but will tolerate partial shade. And although mature trees can tolerate cold temperatures down to 12 degrees Fahrenheit, temperatures below 27 degrees Fahrenheit will kill the blossoms. To plant a tree, select a sunny location that can accommodate the size of a mature tree. Dig a hole at least twice the diameter of the pot the tree came in, about as deep as the container. Place the tree in the hole, loosening or cutting away any girdled roots. The tree should be planted so that the top of the root ball is about one to two inches higher than the soil level. Refill the hole with the original soil. Water the newly planted tree thoroughly.

Mature loquat trees are drought tolerant, but as with any newly planted tree, it requires regular irrigation until it becomes established. During drought years mature trees benefit from irrigation while they are developing fruit to improve the harvest. Trees closely surrounded by a regularly irrigated lawn usually don't require additional watering. Young trees also benefit from a light application of nitrogen fertilizer one to three times a year starting in early spring. Several sources recommend a 6-6-6 NPK (6 percent nitrogen, 6 percent phosphate, 6 percent potassium) fertilizer. Be aware that too much nitrogen reduces blossoming and subsequent fruit production.

Pruning is only required to maintain desired tree size and shape. Pruning is best done after harvest during a dry period to decrease the risk of blight, and before the fall when the tree begins to develop blossoms. Heavy fruiting is common. The loquats are smaller when this happens. When this occurs, you can improve the size of the fruit by hand thinning the blossoms. A major potential disease to loquat trees is fire blight. Fire blight is caused by the bacteria *Erwinia amylovora*. Signs of fire blight include blackened leaves that look like they've been scorched by fire. Immediately remove any leaves and branches with fire blight, seal them in a plastic bag and throw it in the trash. Be sure to disinfect any garden tools used to remove the infected areas in a solution of one part bleach and nine parts water to prevent spreading the bacteria to other plants and trees.

GROWING FROM SEED. Remove the seeds from a fresh ripe loquat. Wash them in water and immediately plant them in a container of well draining soil. Do not let the seeds dry out. The seeds require moist soil and temperatures over 70 degrees Fahrenheit to germinate.

▶ USDA HARDINESS ZONES: 8–11. CHILLING HOURS: 0. WATER REQUIREMENTS: Average.

HOW TO CONSUME. Fresh loquats are best when eaten shortly after harvest. To eat fresh, peel off the skin before consuming since the skin is slightly bitter. Loquats make a nice addition to spring salads. Fresh loquats can also be eaten over cereal. Loquat flesh can be stewed in sugar and eaten as a desert with ice cream. Loquats can be frozen. To freeze them, you need to wash them, and remove the stems at both ends, as well as the seeds. Pack them into a freezer container with a medium sweet syrup (which you can make by dissolving one and three-quarters cup of sugar in four cups of warm water). Make sure you leave at least half an inch of headspace in the container before sealing it and storing it in the freezer. Loquats can also be dried. They should be peeled and pitted before putting them in your electric food dehydrator. Loquats are used in pies, cobbler, and crumbles. Loquats are often used to make loquat jelly.

ADDITIONAL INFORMATION. WARNING: Loquat seeds contain the toxin cyanogenic glycoside. If consumed, the cyanogenic glycoside is broken down and releases hydrogen cyanide which is toxic to both humans and animals. Symptoms of cyanide poisoning include dizziness, confusion, trouble breathing, stomach ache, convulsions and seizures. Immediate medical attention should be sought. Small young tree branches have been used for fodder. In Asia, wood from loquat trees have been used to make walking sticks. Because of its hard wood, it has also been used to make clubs.

REFERENCES

Andress, Elizabeth L., and Judy A. Harrison. *So Easy to Preserve.* 6th ed. Bulletin 989. Athens, GA: College of Family and Consumer Sciences, College of Agricultural and Environmental Sciences, Cooperative Extension, University of Georgia. 2014.

Barth, Brian. "Forgotten Pomes." *Horticulture*, vol. 112, no. 2, March/April 2015, p. 52–57.

Butterfield, Harry. *A History of Subtropical Fruits and Nuts in California.* Berkeley, CA: University of California Division of Agricultural Sciences, Agricultural Extension Service, Agricultural Experiment Station. 1963.

Caballero P., and M.A. Fernandez. "Loquat, Production and Market. In : Llacer G. (ed.), Badenes M.L. (ed.). *First international symposium on loquat.* Zaragoza: CIHEAM, 2003. p. 11–20 (Options Mediterraneenes: Serie A. Seminaires Mediterraneens; n. 58)

Crane, Jonathan H., and M. Lilia Caldeira. "Loquat Growing in the Florida Home Landscape." Institute of Food and Agricultural Services, University of Florida. 1980. https://edis.ifas.ufl.edu/pdffiles/MG/MG05000.pdf. Accessed 01/05/18.

"*Eriobotrya japonica* (Thunb.) Lindl." USDA, NRCS. 2018. The PLANTS Database. National Plant Data Team, Greensboro, NC 27401-4901 USA. https://plants.sc.egov.usda.gov/core/profile?symbol=ERJA3. Accessed 12/21/18.

Ettinger, Jill. "The Lure and Lore of the Loquat." *Organic Authority*. May 10, 2011.

Gilman, Edward F., and Dennis G. Watson. "Eriobotrya japonica: Loquat." University of Florida IFAS Extension. 2015. https://edis.ifas.ufl.edu/st235. Accessed 12/17/15.

LaRue, Ralph G. "Loquat Fact Sheet." UC Davis Fruit & Nut Research & Information. University of California, Davis. http://fruitsandnuts.ucdavis.edu/dsadditions/Loquat_Fact_Sheet/. Accessed 01/01/15.

"Loquat." California Rare Fruit Growers. 1997. www.crfg.org/pubs/ff/loquat.html 02/21/18.

"The Lowdown on Loquats." *The Dominion Post*. January 4, 2014, p. 20.

Morton, Julia F. *Fruits of Warm Climates*. Eugene, OR: Wipf and Stock Publishers, 2003.

Mullins, Lynne. "Eat It Now Loquats." *Sydney Morning Herald*, Nov. 6, 2007. Good Living Section, p. 13.

Ortho All About Citrus & Subtropical Fruits. Des Moines: Meredith Books. 2008.

Pittenger, Dennis. *California Master Gardener Handbook*. 2nd ed. University of California, Agricultural and Natural Resources. 2014.

Popenoe, Wilson. *Manual of Tropical and Subtropical Fruits: Excluding the Banana, Coconut, Pineapple, Citrus Fruits, Olive and Fig*. NY: Macmillan Co., 1920.

Royal Horticultural Society. "(Not-so-exotic) loquats." *Garden*. June 2014. p. 31

Lychee (*Litchi chinensis* Sonn.)

Lychees, also known as lichi, litchi, leechee and lychee nut, are a small, oval shaped, lumpy primarily pink to red skinned fruit, although other colored skinned varieties, such as yellow exist. Measuring from half-inch up to two inches, the lychees' somewhat hard skin is not edible. But the whitish semi translucent fruit pulp is sweet and aromatic. There is a seed in the middle of the fruit. Their taste is commonly described as a blend of a grape and a pear. Lychees only last a few days at room temperature before they begin to deteriorate. But if refrigerated, lychees can last up to five weeks. The lychee is in the same botanic family (*Sapindaceae*) as the longan and rambutan. All three bear whitish semi-translucent fruit pulp.

LYCHEE LORE. During the Tang Dynasty, the Emperor supposedly used fresh lychee to lure his concubine to his palace. Because of her love for the fresh fruit, he had his men ride their fastest horses to deliver to them to the palace from where they grew some 600 miles away. This legend may be why in China the lychee is also regarded as a symbol of romance and love. The Tripuri Tribe of Tripura, northeast India, considers a lychee tree planted near the house to be bad luck.

Lychees are believed to have analgesic, antibacterial, antiviral, anticancer, antitussive, and hypoglycemic properties. In Asian countries lychees have been used to treat a variety of ills ranging from fever, pain, swelling, ulcers, and coughs to diabetes. However, more scientific studies need to be done to examine the safety and effectiveness of using lychees as a treatment for those conditions.

Lychee *(Peggy Greb, USDA Agricultural Research Service/Wikimedia Commons)*

TREE HISTORY. The lychee dates back to 2000 BCE. Originating in the southern China provinces of Fukien and Guangdong, it then spread to other parts of Asia. Lychees made their way from Burma to India near the end of the 17th century. They later made their way to Europe, and then to America. In his document *A History of Subtropical Fruits and Nuts in California,* Harry Butterfield of the University of California reports that the first attempt to grow lychees in California was in 1871. But it wasn't until 1914 that a lychee tree in Santa Barbara successfully produced lychee fruit. Subsequently other trees in San Diego were also reported to bear fruit. A lychee cultivar was first introduced to Hawaii in 1873. Additional cultivars were introduced by the USDA in the early 20th century. Commercial production of lychees in Hawaii reached its high point in the 1960s. Lychees were introduced in Florida from India between 1870 and 1880. But it wasn't until the 1940s that an improved method of propagation spurred production in the state. Commercial production began to take off in the late 1940s.

TREE CHARACTERISTICS. A subtropical and tropical evergreen tree, under ideal conditions lychee trees can grow close to 50 feet tall and equally wide in their native habitat. But in most cases in the United States they generally only grow 20 to 30 feet tall and wide. With a very dense canopy, they are often grown as landscape trees in southern coastal areas. They can also be grown as hedges and in containers. The tree bears terminal clusters up to 30 inches of small yellowish blossoms in the spring. Fruit is borne in clusters. Cross pollination is not required.

Lychee leaves are slightly reddish when young, but as they mature they become dark green and somewhat shiny on the top side of the leaves. The leaves are compound with up to eight leaflets. The trees grow one to three feet per year. They generally don't begin to bear fruit until they are at least three years old. Mature trees can produce upward of 100 pounds of fruit annually.

▶ FAMILY: Sapindaceae. GENUS: *Litchi* Sonn. SPECIES: *Litchi chinensis* Sonn.

SELECTING A TREE. Although approximately 100 varieties of lychee are grown in China and other nations, only about a dozen have been successfully grown in United States. Most of those cultivars bear the red to pink skinned fruits. A commonly grown cultivar is the 'Bengal,' which is often said to be the easiest cultivar to grow.

Other popular medium and large sized seed cultivars include the following: 'Brewster'—produces large bright red fruit with a soft pulp. • 'Emperor'—produces big red fruit. • 'Hak Ip' (a.k.a. 'Hei yeh')—produces medium red fruit with a soft thin skin. • 'Kaimana'—produces large red fruit. • 'Kwa luk'—produces large red fruit with a green tip and line. • 'Mauritius'—produces bright red medium sized fruit. • 'No Mai Tsze'—produces large red fruit supposedly excellent for drying. • 'Sweet Heart'—produces large red fruit. Sample popular small seeded varieties include the following: 'Groff'—produces small red fruit. • 'Sweet Cliff'—produces small pink to red fruit.

Although trees can be grown from seed, successful fruiting varies significantly with trees grown from seed. Most lychee trees are propagated through air layering (a process where a rooting medium is bound to a part of a plant like a branch forcing that part to root so it can later be removed from the plant and put in soil to grow as a new plant). Trees grown through air layering take three to five years to begin bearing fruit. In contrast, trees grown from seed can take up to 10 more years before starting to fruit. The fruit borne by trees grown from seed are said to be of varying quality.

When selecting a tree from the nursery, check the tree's overall health. Look for active growth, such as new or young leaves. Make sure its leaves are not yellow or discolored, and free from insects or signs of insect damage. Check the entire tree to make sure it is free from injury, such as broken branches, cuts in the trunk, etc. And if possible, gently lift the tree out of the pot to check the roots. Overgrown trees that have seriously girdled roots will likely develop a poor root structure so they should be avoided.

GROWING LYCHEES. Lychees have been successfully grown commercially in warm areas free from frost. Young trees are extremely susceptible to frost. Frost also kills any open blossoms. The trees do best in areas with hot summers and dry winters. Remember they are subtropical and tropical trees, but they do need from 100 to 200 hours of chilling time (temperatures between 35 and 45 degrees Fahrenheit). Lychee trees grow in most soils, provided there is adequate drainage and organic matter. They grow best in well drained loamy soil. Poorly drained clay soil without organic matter could lead to root disease.

When planting a lychee tree, choose a sunny location protected from the wind. (High winds can damage new trees.) Dig a hole three to four times wider and deeper than the container in which the tree arrived in. Mix the excavated soil with some organic matter (no more than an equal amount). Water the newly planted tree. Although mature trees are drought tolerant, young trees need regular watering until they become established. Be careful not to over water the tree. Young trees (up to four years old) should be fertilized with a light fertilizer for acid loving plants. Organic fertilizers are preferable since young trees are vulnerable to root burn. Fertilization should be done between March and December. Young trees should be pruned the first few years to develop a strong structure and shape the tree. Mature trees generally don't require pruning. Any pruning of mature fruit bearing trees should be done after harvest. Potential tree pests include scales, aphids, mites, and fruit flies.

HOW TO GROW FROM SEED. To grow from seed, clean away any pulp from the seeds from a fresh lychee. Since the seeds begin to lose viability once removed from the pulp, you should take action to grow the seed the same day. There are a couple methods of germinating the seed. You can plant it in a pot with moist potting soil. Cover the top of the pot with clear plastic and place the pot in indirect sun. Keep the soil warm and moist to keep the seed from drying out. It can take a couple weeks to sprout. The seeds can also be germinated in moist sand in the shade in a warm environment. Another germination method is to soak the seed in water for a few days. Change the water daily. Plant the seed in potting mix when the seed shell begins to crack. Keep the soil moist and warm until the seed sprouts.

▶ USDA HARDINESS ZONES: 9–11. CHILLING HOURS: 100–200 Hours. WATER REQUIREMENTS: Average.

HOW TO CONSUME. Lychees are delicious when eaten fresh. When harvesting lychee, you want to select ripe ones which are firm to the touch but slightly soft when squeezed. Lychees soft to the touch are over ripe. Lychees should only be harvested ripe since they do not ripen once picked from the tree. They should be stored in the refrigerator. Peel off the outer shell using your finger nail or a knife. Or you can open the shell at one end of the fruit and squeeze out the pulp. Remember there is an inedible seed or in some cases seeds in the center. Lychee varieties with a small seed in the center tend to sell at a higher price.

Fresh lychees can be added to salads and to desserts. They can also be added to beverages like tea for added flavoring. Lychees are frequently commercially canned. If you're unable to find fresh lychees, you can find canned lychees for sale in specialty and Asian supermarkets. Lychees can be use to make sauces, jams and other preserves. Lychee sorbet and ice cream are also served in select restaurants. Lychees are frequently used to make alcoholic cocktails like lychee martinis, lychee cosmopolitans and lychee daiquiris. They are also used to make other alcoholic beverages like lychee wine and lychee beer coolers. Lychees are frequently dried. Research on drying of peeled, thin sliced lychee showed little change in the dried lychees' light brown color when dried from 50 to 70 degrees Celsius.

ADDITIONAL INFORMATION. There have been reported outbreaks of hypoglycemic encephalopathy (a brain disorder or disease) in India and Vietnam in children associated with consumption of lychees. An article in the medical journal *Lancet* explains that lychee contain "unusual amino acids that disrupt gluconeogenesis and β-oxidation of fatty acids." The neuro toxicity of lychees in children depends upon the number of raw lychees consumed, the concentration of hypoglycemic acid, the child's age and nourishment status. The cost of fresh lychees in the United States and therefore a lower consumption of them and the resulting overall nourishment of children most likely account for no lychee related encephalopathy outbreaks in America. A fruit, lychees are sometimes referred to as lychee nuts. This is most likely because lychees allowed to dry on the trees have the appearance of a nut. One cup of fresh lychees is approximately 125 calories. In the United States, Florida produces the most lychees, followed by Hawaii and then California. Worldwide, China is the largest producer of lychees, followed by India.

REFERENCES

Gilman, Edward F., and Dennis G. Watson. *Litchi chinensis:* Lychee. ENH-523. University of Florida, IFAS Extension. https://edis.ifas.ufl.edu/pdffiles/ST/ST36400.pdf. Accessed 02/07/16.

Ibrahim, Sabrin R.M., and Gamal A. Mohamed. "*Litchi chinensis:* Medical Uses, Phytochemistry, and Pharmacology." *Journal of Ethnopharmacology.* Vol. 174, no. 4, Nov. 2015, p. 492–513.

Janjai, S., et al. "Thin-layer Drying of Litchi (*Litchi chinensis* Sonn.)" *Food and Bioproducts Processing,* v. 89, no. 3, July 2011, p. 194–201.

"*Litchi chinensis* Sonnerat, Voy." *Flora of China,* v. 12, p. 6. www.efloras.org/florataxon.aspx?flora_id=2&taxon_id=200013205. Accessed 03/09/18.

"Litchi (Lychee) Production Guide." *Business Diary (Philippines)* Nov. 29, 2017. http://business diary.com.ph/8242/litchi-lychee-production-guide/. Accessed 02/07/16.

"Lychee." California Rare Fruit Growers. www.crfg.org/pubs/ff/lychee.html. Accessed 10/22/15.

Menzel, Christopher. *Lychee, Its Origin, Distribution and Production Around the World.* The Archives of the Rare Fruit Council of Australia. March 1995. http://rfcarchives.org.au/ Next/Fruits/Litchi/WorldLychees3-95.htm. Accessed 03/10/18.

Mitra, S.K. *Overview of Lychee Production in the Asia-Pacific Region. RAP Publication 2002/04.* Food and Agricultural Organization of the United Nations, Regional Office for Asia and the Pacific, Bangkok, Thailand. March 2002. www.fao.org/docrep/005/ac684e/ac684e04. htm#bm04. Accessed 03/10/18.

Morton, Julia F. *Fruits of Warm Climates.* Eugene, OR: Wipf and Stock Publishers, 2003.

Mullins, Lynne. "Best Eaten in the Buff." *The Newcastle Herald.* February 1, 2012, p. 44.

Nagy, Steven, and Philip E. Shaw. *Tropical and Subtropical Fruits: Composition, Properties and Uses.* Westport, CT: AVI Publishing. 1980.

Ortho All About Citrus & Subtropical Fruits. Des Moines, IA: Meredith Books. 2008.

Popenoe, Wilson. *Manual of Tropical and Subtropical Fruits: Excluding the Banana, Coconut, Pineapple, Citrus Fruits, Olive and Fig.* NY: Macmillan Co., 1920.

Prasad, J.S., Raj Kumar, Mukund Mishra, et al. "Characteristics of Litchi Seed Germination." *HortScience,* v. 31, no. 7, Dec. 1996, p. 1187–1189.

"SelecTree: Litchi chinensis Tree Record." California Polytechnic University. https://selectree. calpoly.edu/tree-detail/litchi-chinensis. Accessed 03/08/18.

Sharma, M., C.L. Sharma, and J. Debarrma. "Ethnobotanical Studies of Some Plants Used by Tripuri Tribe of Tripura, NE India with Special Reference to Magico Religious Beliefs." *International Journal of Plant, Animal and Environmental Sciences,* v. 4 #3, July–Sept. 2014, p. 518–528.

Spencer, Peter S., and Valerie S. Palmer. "The Enigma of Litchi Toxicity: An Emerging Health Concern in Southern Asia." *Lancet,* v. 5, April 2017.

Takele, Etaferahu. *Projected Costs to Establish a Lychee Orchard and Produce Lychees. Coastal Regions of California, 2002.* University of California Cooperative Extension. Reprinted 2005. https://coststudyfiles.ucdavis.edu/uploads/cs_public/4b/ac/4baca531-5bd2-4877-8179-63d706657905/lycheescc02.pdf. Accessed 02/06/16.

Vieth, Robert. *Lychee.* University of California Cooperative Extension, Ventura County. http://ceventura.ucanr.edu/Com_Ag/Subtropical/Minor_Subtropicals/Lychee/. Accessed 02/06/16.

Wood, Marsha. "Luscious Lychee: Scientists Find Keys to Plentiful, Predictable Harvests." *Agricultural Research,* May/June 2009, p. 21.

Wood, Marsha. "New Options for Lychee and Longan Fans and Farmers." *Agricultural Research,* May 2004, p. 20–22.

Zee, Francis, Mike Nagao, Melvin Nishina, and Andrew Kawabata. *Growing Lychee in Hawaii. F&N-2.* University of Hawaii at Manoa Cooperative Extension, College of Tropical Agriculture and Human Resources. June 1999. www.ctahr.hawaii.edu/oc/freepubs/pdf/F_N-2. pdf. Accessed 03/22/18.

Malay Apple (*Syzygium malaccense* [L.] Merr. & L.M. Perry)

Also known as a mountain apple, water apple, Otaheite apple, wax jambu, French cashew, and Malaysian apple, the fruit is shaped like a bell and is generally two to four inches long and about two and a half inches wide. Immature Malay apples bear a smooth thin white skin. As they mature, the skin becomes waxy and changes in color from white to red or crimson or purple, or may remain white with streaks of pink and or red depending upon the cultivar.

However, most Malay apples bear a red skin. The pulp is white, spongy or crisp. Each apple contains one or two brown seeds about an inch in diameter, although some trees bear fruit with no seeds. The Malay apple has a very mild, sweet apple taste with a slight rose fragrance. But some people consider its taste and texture to be more like a Comice pear than an apple. Fresh Malay apples have a very short shelf life of about only four days once picked from the tree. The apples also bruise very easily.

Although somewhat similar in appearance and taste, the Malay apple is not the same as a Java apple. Both the Java apple and Malay apple are in the same botanic family and genus, but are different species.

MALAY APPLE LORE. Malay apple trees play a minor role in a number of folk tales and myths. For example, in one myth from Melanesia two men climb a Malay apple tree to eat the fruit. But an apple drops from the tree, killing a pig below the tree. When the men climb back down the tree, they discover the dead pig belongs to an ogress. Out of fear, they climb back up the tree to escape her ire. The ogress uses her own intestines to help her climb up the tree so she can eat the men. But the men smartly cut her intestines and she falls to her death. And in a folktale from New Guinea, Malay apple

Malay apple *(McKay Savage from London, UK/Wikimedia Commons)*

trees grow by a river in tribute to a woman who was born from a Malay apple. In Hawaii and Fiji, natives used to offer Malay tree blossoms to Pele, the volcano goddess.

Various parts of the Malay apple tree have been used in folk medicine in other countries where the tree grows. It's been used to treat a variety of conditions ranging from coughs to diabetes. The tree leaves reportedly prevent inflammation, and the tree bark has been used to treat ailments from mouth infections to sexually transmitted diseases.

TREE HISTORY. Originating in Malaysia, it spread to the Pacific Islands in ancient times by Polynesian settlers from Malaysia. It spread to India and Africa. The tree is recorded in Hawaii prior to the arrival of the missionaries. From the Pacific Islands they were brought to Jamaica in 1793. The tree was documented in the United States in 1839 and in St. Croix and Bermuda in the 1870s. By the 1920s the trees were in South and Central America.

TREE CHARACTERISTICS. A very pretty flowering tree, the evergreen Malay apple tree grows 30 to 60 feet tall, and has an upright trunk measuring up to 15 feet in circumference. A fairly fast growing tree, it bears either a pyramid shaped or cylindrical crown. It has long (six to 18 inches and four to eight inches wide) evergreen leathery leaves.

The trees beautiful blossoms have led many to consider the Malay apple tree one of the most attractive flowering tropical fruit trees. There can be two or three flowering seasons in a year. It's commonly thought that flowering is prompted by wet weather following a dry spell. Mature trees can produce approximately 50 to 200 pounds of apples per year.

The blossoms have only a mild fragrance, but are large (two to three inches wide) and colorful. In shades of pink to red, the blossoms are borne in short stalked clusters of up to eight blossoms, on the upper tree trunk, and also cover leafless sections of mature branches. The blossoms are pollinated primarily by bees. After a few days the blossoms fall to the ground creating a gorgeous colorful carpet of blossoms. Thus the trees are grown more frequently as landscape trees than as commercial fruit trees. The trees are primarily propagated by air layering. But they are also propagated by seed, and by grafting or budding on same species rootstock.

▶ FAMILY: Myrtaceae. GENUS: *Syzygium* P. Br. ex Gaertn. SPECIES: *Syzygium malaccense* (L.) Merr. & L.M. Perry.

SELECTING A TREE. There are a number of different Malay apple cultivars, including at least one that is flowering only. Make sure when you select a tree that it is a fruiting cultivar if you want Malay apples. When selecting a tree

Malay apple blossoms *(Tau'olunga/Wikimedia Commons)*

from the nursery, check the tree's overall health. Look for active growth, such as new or young leaves. Make sure its leaves are not yellow or discolored, and free from insects or signs of insect damage. Check the entire tree to make sure it is free from injury, such as broken branches, cuts in the trunk, etc. And if possible, gently lift the tree out of the pot to check the roots. Overgrown trees that have seriously girdled roots will likely develop a poor root structure so they should be avoided.

How to Grow a Malay Apple Tree. The tree is a tropical tree that needs a humid climate. Frost can easily kill young trees. The trees grow in a wide range of soils from sandy to clay soils. But the tree does not grow well in high alkaline soils.

To plant a tree, select a sunny location that can accommodate the size of a mature tree. Dig a hole at least twice the diameter of the pot the tree came in, about as deep as the container. Place the tree in the hole, loosening or cutting away any girdled roots. The tree should be planted so that the top of the root ball is about one to two inches higher than the soil level. Refill the hole with the original soil. Water the newly planted tree thoroughly. The tree should be watered regularly until it becomes established. Pruning is only needed to remove dead branches, to allow more light in the tree, and to shape it as necessary.

How to Grow from Seed. Malay apple seeds are only viable for a short period after removal from the fruit. So they should be planted shortly after they are secured. Clean and wash away any pulp from the seed. Plant it in moist soil about half-inch deep. Keep the soil moist. It should germinate within six weeks. Trees grown from seed usually take four to six years before they begin bearing fruit.

▶ USDA Hardiness Zones: 10–11. Chilling Hours: 0. Water Requirements: Minimum of 60 inches of rainfall annually; also needs watering during dry periods.

How to Consume. Like other apples, the Malay apple can be eaten fresh. Fresh apples slices can be added to salads, dipped into chocolate or caramel, or added to smoothies. But because it has such a mild taste, many consider it almost completely lacking in flavor. So the apple is frequently stewed, or otherwise cooked with other ingredients. Malay apples are frequently pickled, used in jams, jellies and chutneys. Wine has been made from Malay apples. In Indonesia the blossoms are preserved in syrup, or added to salads. Indonesians have been known to eat young leaves cooked like greens.

Additional Information. Although the tree is grown both as a landscape tree, and as a fruit tree, it is generally not grown as a commercial fruit tree, because of the fruits' easy bruising and short shelf life. The Malay apple tree wood was carved by ancient Hawaiians to make statues and idols. The tree wood has also been used for construction. The inner bark of the tree has been used to make a brown dye.

References

"About Malay Apple Trees." Botanical Growers Network, Gainesville, FL. http://botanical growersnetwork.net/znetsol/1-ProductPages/T1g/SYZ-MAL/pop.htm. Accessed 10/15/18.

Joyner, Gene. "The Malay Apple." The Archives of the Rare Fruit Council of Australia. March 1994. http://rfcarchives.org.au/Next/Fruits/MyrtaceaeFamily/MalayApple3-94.htm. Accessed 10/16/18.

Khoo, Hock Eng, et al. "Phytochemicals and Medicinal Properties of Indigenous Tropical Fruits with Potential for Commercial Development." *Evidence-Based Complementary and Alternative Medicine*, vol. 2016, p. 1–20. www.ncbi.nlm.nih.gov/pmc/articles/PMC4906201/. Accessed 11/02/18.

"Malay Apple." Cooks Info. www.cooksinfo.com/malay-apples. Accessed 10/15/18.

Morton, Julia F. *Fruits of Warm Climates.* Eugene, OR: Wipf and Stock Publishers, 2003.

"Mountain Apple Care: Tips for Growing Mountain Apple Trees." Gardening Know How. www.gardeningknowhow.com/ornamental/trees/mountain-apple/mountain-apple-information.htm. Accessed 10/16/18.

Orwa, C., et al. "Syzygium malaccense." 2009 Agroforestree Database: A Tree Reference and Selection Guide Version 4.0. www.worldagroforestry.org/treedb/AFTPDFS/Syzygium_malaccense.PDF. Accessed 10/14/18.

Popenoe, Wilson. *Manual of Tropical and Subtropical Fruits: Excluding the Banana, Coconut, Pineapple, Citrus Fruits, Olive and Fig.* NY: Macmillan Co., 1920.

Riesenfeld, Alphonse. *The Megalithic Culture of Melanesia.* Leiden: E.J. Brill, 1950.

"Syzygium malaccense." Tropical Plants Database. http://tropical.theferns.info/viewtropical.php?id=Syzygium+malaccense. Accessed 10/16/18.

"*Syzygium malaccense* (L.) Merr. & L.M. Perry." US Department of Agriculture—NRCS. The PLANTS Database. National Plant Data Team, Greensboro, NC. http://plants.usda.gov. Accessed 10/14/18.

"Syzygium malaccense—Mountain Apple." Hawaiian Plants and Tropical Flowers. https://wildlifeofhawaii.com/flowers/1152/syzygium-malaccense-mountain-apple/. Accessed 10/14/18.

Watson, Molly. "Ohi'a 'ai (Hawaii Mountain Apples)." The Spruce Eats. www.thespruceeats.com/hawaii-mountain-apples-4150745. Accessed 10/16/18.

Mandarin Orange (*Citrus reticulata* Blanco)

The name Mandarin is used to refer to a broad group of oranges with a deep orange thin skin that are easy to peel. Many people confuse a tangerine as being the same as a mandarin orange, but they are not the same. A tangerine is technically a variety of the mandarin orange. Mandarin oranges are shaped like a flattened sphere. The thin skin peels easily away from the fruit's sweet juicy pulp. The pulp does contain a few small inedible seeds.

MANDARIN LORE. During Chinese New Year you'll see lots of Mandarin orange trees showing up in homes and businesses. They symbolize good luck and good fortune, and the more fruit on the tree, the more luck it will bring. The Chinese word for mandarin orange sounds a lot like the Chinese word for gold. Plus, the mandarin looks like the sun, which is aligned with the yang (positive) principle. Mandarin oranges are a common gift during the holidays and New Years and are given as a token of good fortune and luck. It's preferred to gift mandarin oranges with the stem and a few leaves attached. Leaves and stems attached to the mandarin represent fertility and long life.

There are claims being made that mandarin oranges can cure asthma, indigestion, clogged arteries, irritable bowel syndrome, and liver disease. But there's currently insufficient research to support these claims.

TREE HISTORY. The Mandarin orange originated in Asia. Some believe it originated in China, hence the name Mandarin orange. But others believe it originated in Southeast Asia and the Philippines. They were introduced into England from China in 1805, and from there, they were subsequently introduced into the Mediterranean. By 1850 the trees were well established in Italy. They were imported into the U.S. between1840 and 1850 by the Italian Consul

Mandarin orange *(Pxhere)*

in New Orleans. And in 1896 a variety from Japan was also introduced into the United States.

Mandarin cultivars are numerous and are generally separated into four groups: Satsuma, Tangerine or Common, King, and Mediterranean. Numerous cultivars of Satsumas originated in Japan where they are a prominent citrus fruit. Satsumas are almost seedless and bear a reddish orange peel. Some Satsuma cultivars include the 'Owari,' 'Obawase,' 'Okitsu,' 'Kara,' and 'Kimbrough.'

Tangerines are an important group of mandarins because of their widespread consumption and cultivation. There are dozens of varieties of tangerines. But probably the most popular is the Clementine, which originated in Algeria and was brought to the United States in 1909. King mandarins are mandarins from IndoChina of importance because they are the parents of cultivars such as the 'Kinnow' and 'Encore.'

Mediterraneans are mandarin cultivars that originated in the Mediterranean basin. A popular cultivar is the 'Willowleaf.' It's a medium sized mandarin with a yellowish orange peel and flesh.

There's also a wide variety of Mandarin hybrids. The two most popularly known Mandarin hybrids are the tangor and the tangelo. The tangor is a cross between a tangerine and an orange. A tangelo is a cross between a tangerine and a grapefruit.

Numerous mandarin cultivars are constantly being released. Of all the citrus, mandarin cultivar introductions are the most numerous. In the 15 year period from 1986 to 2001 alone, 62 new cultivars were released in the world.

TREE CHARACTERISTICS. Like other citrus trees, Mandarins are evergreen trees. Its dark green glossy leaves can be broad or slender, and are shaped like a lance head. Among citrus trees, they are fairly cold hardy, second only to the kumquat. They are small to medium sized trees that have a spreading canopy and thorny branches. They bear fragrant blossoms. The trees tend to be alternate bearing, meaning they tend to bear a bountiful crop one year, followed by a minimal crop the subsequent year.

Although the trees can be grown from seed, most trees are propagated by bud grafting. Trees propagated by seed may not bear true to type, and also take a long time before beginning to bear fruit. Grafted trees generally begin bearing fruit in a couple of years.

▶ FAMILY: Rutaceae. GENUS: *Citrus* L. SPECIES: *Citrus reticulata* Blanco.

SELECTING A TREE. There are many varieties and cultivars of mandarins, many with very distinct tastes. Be sure you enjoy the taste of the specific mandarin cultivar tree you select for purchase. Sample popular cultivars include the following: 'Clementine'—produces juicy easy to peel fruit with few seeds. • 'Fairchild'—produces sweet and juicy fruit but has many of seeds. • 'Honey'—produces small sweet fruit but has many seeds, • 'Kinnow'—produces juicy fruit with seeds. • 'Willowleaf'—produces sweet, juicy, aromatic fruit.

When selecting a tree from the nursery, check the tree's overall health. Look for active growth, such as new or young leaves. Make sure its leaves are not yellow or discolored, and free from insects or signs of insect damage. Check the entire tree to make sure it is free from injury, such as broken branches, cuts in the trunk, etc. And if possible, gently lift the tree out of the pot to check the roots. Overgrown trees that have seriously girdled roots will have a poor root structure so they should be avoided.

HOW TO GROW A MANDARIN TREE. Mandarin trees grow in most soils, providing they provide adequate drainage. The trees do best in soil with a pH between 6.0–6.5. They prefer full sun.

The best time to plant a tree is in the spring when all danger of frost has passed. Planting in the spring allows the tree some time to become established before winter. To plant a tree, select a sunny location or one with partial shade that can accommodate the size of a mature tree. Dig a hole at least twice the diameter of the pot the tree came in, about as deep as the container. Place the tree in the hole, loosening or cutting away any girdled roots. The tree should be planted so that the top of the root ball is about one to two inches higher than the soil level. Refill the hole with the original soil. Water the newly planted tree thoroughly. The tree should be regularly watered the first two years.

Young citrus trees are sensitive to frost. If temperatures are expected to dip to 29 degrees Fahrenheit or lower, you need to protect the tree from frost. You should wrap the tree trunk and branches in insulation material like cardboard, fiberglass or frost protection covers. Make sure the soil is moist and not dry, since damp soil retains and radiates more heat than dry soil. Bare soil also radiates more heat than soil covered with mulch or other ground covers.

If your tree does sustain frost damage, don't prune away any dead branches until the spring. This allows time for the tree to recover in warmer weather. It also allows you to better identify the damaged branches to be removed. Young citrus trees benefit from fertilization. If using a citrus fertilizer, follow the instructions on the label. Pruning is only required to control the size of the tree, and to increase light and airflow through the tree canopy. Therefore only minimal pruning is required.

How to Grow from Seed. To grow a tree from seed, remove the seeds from the mandarin. Wash any pulp away from the seeds. Soak the seeds in clean water overnight. Place the seeds about half-inch deep in a sterile moist planting/potting medium. Cover the pot with plastic to create a humid atmosphere. Keep the soil moist until seedlings sprout. The seeds germinate best in temperatures above 60 degrees Fahrenheit. The seeds usually germinate in two to three weeks, but can take up to three months. Once the seeds sprout, transplant them into a larger pot and move them to a sheltered location outdoors. Be aware that trees grown from seed do not always bear true to type. And trees grown from seed may take up to 10 years before bearing fruit.

▶ USDA Hardiness Zones: 8–11. Chilling Hours: 0. Water Requirements: Average.

How to Consume. Mandarins are usually eaten fresh. To consume fresh, wash and rinse the mandarin under running water and pat dry the mandarin with a paper towel. Then peel away the outer skin. Eat the segmented pulp, being careful not to eat any seeds in the fruit. Fresh mandarin segments are frequently added to salads. The juice is used to help flavor or season meats and poultry. The pulp and juice is also used in baked goods, such as cake and breads. Mandarins are sometimes used in marmalade and jam recipes. They can also be juiced. The citrus is used in mixed alcoholic beverages, and the peel can be used to flavor liquors.

Additional Information. Fresh mandarins can be stored three or more weeks in the refrigerator (38 to 48 degrees Fahrenheit). China produces the most mandarins, approximately 21.2 million tons, more than 70 percent of worldwide production. Other major producers include the European Union

(3.1 million tons), Turkey (1.3 million tons), and Morocco (1.2 million tons). Production of mandarins in the United States is approximately 758,000 tons annually.

REFERENCES

Andersen, Peter C., and James J. Ferguson. "The Satsuma Mandarin." HS-195. University of Florida, Institute of Food and Agricultural Sciences, Cooperative Extension. November 2015. http://edis.ifas.ufl.edu/ch116. Accessed 08/19/18.

China/FAO Citrus Symposium Proceedings, May 14–17, 2001, Beijing, People's Republic of China. United Nations Food and Agriculture Organization. http://www.fao.org/docrep/003/x6732e/x6732e12.htm#l. Accessed 08/18/18.

"Citrus—Mandarin, Tangerine." LEAF Network, Az. https://leafnetworkaz.org/resources/PLANT%20PROFILES/Citrus_Mandarin_profile.pdf. Accessed 08/24/18.

"*Citrus Reticulata* Blanco." USDA, NRCS. 2018. The PLANTS Database. National Plant Data Team, Greensboro, NC 27401-4901 USA. https://plants.sc.egov.usda.gov/core/profile?symbol=CIRE3. Accessed 07/03/18.

"Citrus: World Markets and Trade." U.S. Department of Agriculture, Foreign Agricultural Service, July 2018. https://apps.fas.usda.gov/psdonline/circulars/citrus.pdf. Accessed 08/26/18.

Cook, Steve. "Why Are Mandarin Oranges a Symbol of Good Fortune?" S&J Mandarin Grove. September 14, 2016. www.sandjmandarins.com/mandarin-oranges-symbol-good-fortune. Accessed 08/24/18.

Fake, Cindy. "A Mandarin by Any Other Name." Publication #31-111. University of California, Agriculture and Natural Resources, Cooperative Extension, Placer/Nevada Counties, December 2004. http://ucanr.edu/sites/placernevadasmallfarms/files/211585.pdf. Accessed 08/26/18.

Geisel, Pamela M., and Carolyn L. Unruh. *Frost Protection for Citrus and Other Subtropicals.* ANR Publication 8100. University of California, Division of Agriculture and Natural Resources, 2003.

Grimes, Julianna. "Mandarin Oranges." *Cooking Light,* vol. 24, no. 11, December 2010, p. 23.

"Mandarins." University of California, Cooperative Extension, Horticulture and Small Farms, Placer County. http://ucanr.edu/sites/ceplacerhorticulture/EatLocal/SignatureCrops/Mandarins/. Accessed 08/18/18.

"Mandarins, Tangerines & Clementines." Produce Blue Book. www.producebluebook.com/wp-content/uploads/KYC/Mandarin-Orange-Family.pdf. Accessed 08/20/18.

Morton, Julia F. *Fruits of Warm Climates.* Eugene, OR: Wipf and Stock Publishers, 2003.

Nolte, Kurt. "Mandarins." College of Agriculture and Life Sciences, University of Arizona. https://cals.arizona.edu/fps/sites/cals.arizona.edu.fps/files/cotw/Mandarin.pdf. Accessed 08/21/18.

Ortho All About Citrus and Subtropical Fruits. Des Moines, IA: Meredith Books, 2008.

Peters, Geoff. "Is It a Mandarin or Tangerine?" Pennsylvania State University Extension. https://extension.psu.edu/is-it-a-mandarin-or-tangerine. Accessed 08/19/18.

"Sample Costs to Establish a Mandarin Orchard and Produce Mandarins (Satsuma)." MD-IR-08. Intermountain—Sierra Nevada Foothills, Placer/Nevada Counties. University of California, Cooperative Extension, 2008. http://ucanr.edu/sites/placernevadasmallfarms/files/142560.pdf. Accessed 08/18/18.

Sauls, Julian. "Home Fruit Production—Mandarins." Texas A&M University. December 1998. https://aggie-horticulture.tamu.edu/citrus/mandarins.htm. Accessed 08/19/18.

"Tangerine." WebMd. www.webmd.com/vitamins/ai/ingredientmono-1515/tangerine. Accessed 08/21/18.

Thulaja, Naidu Ratnala. "Mandarin Orange." National Library Board, Singapore. http://eresources.nlb.gov.sg/infopedia/articles/SIP_205_2005-01-28.html. Accessed 08/19/18.

Yu, Andrea. "Why Mandarin Oranges Mean Luck in the New Year." MyRecipes.com. January 27, 2017. www.myrecipes.com/extracrispy/why-mandarin-oranges-mean-luck-in-the-new-year. Accessed 08/19/18.

Mango (*Mangifera indica* L.)

The mango fruit, known by many other names such as the manga, mangot, manako, and meneke, is frequently oblong, but can also be round, oval, or slightly kidney shaped. They can measure up to 10 inches long, and weigh anywhere from eight ounces up to almost two pounds. Its skin is smooth, leathery, waxy and somewhat thick. The skin color varies widely from different shades of green, to different shades of yellow, and red, and many have color blushes ranging from purple, red, pink and yellow to almost white. The skin is not edible and contains a sap that can cause an allergic rash like poison ivy in sensitive people.

The fruit's pulp ranges in color from a light yellow to a deep orange and is juicy. The pulp is like that of a peach but can be minimally to heavily fibrous. The most desirable commercially produced mangos are those that are minimally fibrous. In the center of the pulp is a flattened or oval shaped stone (although a seedless mango cultivar has been developed in India). Inside the stone is a starchy seed. Mangos can have a pleasant fragrance, but some can have a turpentine-like odor or flavor. Those lacking the turpentine odor are the most desirous. Mango flavors can range from sweet, to sub-acid, to tart.

MANGO LORE. Because mangos date back to ancient times and are grown throughout tropical areas, there are a multitude of myths pertaining to them. One concerns the daughter of the Sun God who was said to be so beautiful, a powerful king falls in love with her and marries her. A powerful sorceress, or in other versions the king's first wife, is jealous and decides to terrify the young woman. To escape the sorceress, she dives into a pool of water and is transformed into a lotus flower. Upon seeing the lotus flower, the king falls in love with its beauty. Angry that even as a flower the king adores its beauty, the evil woman sets fire to the lotus flower. But out of the ashes grew a mango tree. When the first mango fruit fell from the tree, the daughter of the sun burst out of its flesh. Recognizing his wife reincarnated, the king took her home with him and they lived happily ever after.

There's another romantic myth, a Romeo and Juliet type story from the Philippines. The story centers around Pangga, a smart, kind, multi-talented, young woman, and a man named Manong. Manong was a homeless poet who all the women fell for because of his sweet talk. But he only had eyes for Pangga. He promised to bring down the sun and moon to shine on her home. Pangga fell in love with Manong. However, Pangga's parents didn't think Manong was good enough for their daughter and forbid them from marrying. But Pangga and Manong pledged to love one another until death. One day the two of them ran into the woods, never to be seen again. Subsequently a new

Mango *(Renee Comet, National Cancer Institute)*

kind of tree was discovered. The tree's fruit is somewhat crescent shaped like the moon, yellow like the sun, and its taste is sweet like Manong's tongue. It was rich in nutrition like Pangga's multifaceted talents. Believed to be the magical offspring of Manong and Pangga, it was dubbed a 'Manga,' a mix of their two names. Today the manga is known as the mango.

Because Buddha was said to frequently rest in mango groves, mangos are considered sacred to Buddhists. They would carry mangos with them on long expeditions, thereby spreading and popularizing the fruit. In ancient India, rulers were said to plant mango trees along roadsides as a symbol of prosperity. Supposedly Alexander the Great was a fan of the mango and carried them home with him back to Greece.

TREE HISTORY. Mango trees are native to southern Asia, especially Burma and eastern India. In the fourth and fifth centuries BCE, Buddhist monks are believed to have carried the mango to Malaya and eastern Asia, with the Persians carrying them to East African in the tenth century CE. It made its way into the East Indies. It was introduced into the Philippines in the 15th century. In the 16th century it made its way to West Africa and Brazil, then to the West Indies. In the 19th century it landed in Europe and Mexico. It made its way into Florida in 1833 and into California prior to 1857. The U.S. Department of Agriculture officially made 528 mango introductions into the U.S.

from India and other sources from 1899–1937. Mangos now grow in all the tropical areas of the world.

TREE CHARACTERISTICS. The mango tree is a subtropical to tropical evergreen tree that can grow up to 100 feet tall, but in the U.S. it only grows about 30 to 50 feet tall with a canopy up to 30 feet wide. The tree has a deep taproot. In deep soil, the taproot grows to a depth of 20 feet. It also has a wide spreading feeder root system that grows several feet deep. It features a wide, round, rather dense canopy. The tree features large, leathery, oval shaped alternate leaves, usually measuring about 12 inches long but can measure from four up to 16 inches long. New leaves are yellowish green or red wine colored, but turn dark green as they mature, with the bottom of the leaf being a lighter shade of green.

The tree is self fruitful and produces blossom clusters measuring up to 16 inches long. The blossom clusters have thousands of white or yellowish white or pink to red flowers. But a small number of the blossoms are imperfect and not able to produce pollen so they're not able to produce fruit. The blossoms are mildly fragrant and pollinated by insects. In addition to insects, successful pollination also requires temperatures above 55 degrees Fahrenheit and low humidity since the blossom pollen isn't shed in high humidity or rain. Mature trees are alternate bearing. The trees are considered fast growing, growing two feet or more per year. They also have a long life span. Some trees in Asian are reportedly 300 years old and still blossoming. However, trees in the United States are not recorded as having so long a lifespan.

Over 500 mango varieties are said to exist. There are hundreds of cultivars with new cultivars constantly being developed. Mango trees are primarily propagated by grafting and by seed. To a much lesser extent, they are grafted by air layering and budding.

▶ FAMILY: Anacardiaceae. GENUS: *Mangifera* L. SPECIES: *Mangifera indica* L.

SELECTING A TREE. Although there are numerous cultivars, different cultivars are usually recommended for different states. If you live in California, a sample of frequently recommended cultivars include the following: 'Aloha'—produces red, sweet and almost fiberless fruit. • 'Cooper'—produces green fruit with high quality pulp. • 'Costa Rica'—produces pale green, juicy fruit. • 'Edgehill'—produces green fruit with a red blush, is almost fiberless. • 'Manila'—produces yellow fiberless fruit with a sharp flavor. • 'Ott'—produces orange yellow fruit with a pink blush. • 'Reliable'—produces yellow fruit with a red blush, and is almost fiberless. • 'Surprise'—also produces yellow fruit with red blush, and is almost fiberless. • 'T1'—produces green fiberless fruit with a red blush. • 'Thomson'—produces yellow sweet fruit. This cultivar also resists mildew. • 'Villaseñor'—produces greenish yellow fruit with a pink blush and mild flavor.

Sample frequently recommended cultivars for Florida include the following: 'Carabao'—bears greenish yellow fruit that can also be eaten while green but firm. • 'Edward'—produces fiberless yellowish green fruit with a red blush. • 'Florigon'—bears yellow, fiberless fruit. • 'Haden'—produces yellow red, and green fruit. • 'Keitt'—produces green fruit with a rich flavor. This cultivar is also mildew resistant. • 'Kent'—produces fiberless greenish yellow fruit with a red blush. • 'Sensation'—bears yellow fruit with a red blush and is almost fiberless. • 'Tommy Atkins'—bears orange yellow fruit with a red blush and medium fiber.

For Hawaii, sample recommended cultivars include the following: 'Rapoza'—bears yellow fiberless fruit with a red or purple blush. • 'Pirie'— bears yellowish green fiberless fruit. • 'Haden'—produces yellow red, and green fruit with a good flavor. • 'Ah Ping'—produces yellow and orange fiberless fruit. • 'Gouviea'—bears light green with red and yellow blotches fruit that's almost fiberless. • 'Pope'—produces greenish yellow fruit.

When selecting a tree from the nursery, check the tree's overall health. Look for active growth, such as new or young leaves. Make sure its leaves are free from insects or signs of insect damage. Check the entire tree to make sure it is free from injury, such as broken branches, cuts in the bark, etc.

How to Grow a Mango Tree. The most important factor in successfully growing a mango tree is the climate. Mango trees require a frost-free climate since both the tree blossoms and fruit can be killed when the temperature drops below 40 degrees Fahrenheit. The trees fruit best when they have four months of rainfall followed by eight months of dry weather. The trees grow in a wide range of soils from sandy loam soils to clay soils provided they are well draining. Deep rich soils are preferred because they allow for the development of deep taproots to help steady them in high winds and to help them during drought periods. The trees prefer a soil pH of 5.5 to 7.5.

The trees require full sun. To plant a tree, select a sunny location sheltered from the wind. (Wind can damage the blossoms and reduce fruit yield.) Dig a hole at least twice the diameter of the pot the tree came in, about as deep as the container. Place the tree in the hole, loosening and spreading out any girdled roots. The tree should be planted so that the top of the root ball is about one to two inches higher than the soil level. Refill the hole with the original soil. Water the newly planted tree thoroughly.

Young trees benefit from the application of an NPK fertilizer with a ratio of 1:1:1 every three months until the tree begins to fruit. Once the tree begins to fruit, fertilization is generally no longer necessary. Pruning is only needed to control tree size. However, pruning can also help stimulate growth. It's also thought that pruning away some blossom clusters during heavy bloom years may decrease alternate fruit bearing. Common pests include mealy bugs, scales, mites and thirps. The trees are also susceptible to anthracnose disease and powdery mildew.

How to Grow from Seed. To grow a mango tree from seed, remove the stone from a fully ripe fresh mango, preferably harvested from a tree, and not one purchased at the supermarket. Remove any pulp around the stone. Then, using a knife, carefully open the stone husk, being careful not to cut the seed inside. Place the seed on its concave edge in moist potting mix with a quarter of the seed protruding above the soil. Keep the soil moist. The seed should sprout within two weeks in warm tropical climates, and within three weeks in cooler climates. Bear in mind that trees grown from seed may not bear true to type. Also, mango trees grown from seed may take six years before bearing any fruit.

▶ USDA HARDINESS ZONES: 10–11. CHILLING HOURS: 0. WATER REQUIREMENTS: 30–100 inches rainfall annually.

How to Consume. Mangos are enjoyed fresh. To eat a fresh mango, first select a ripe one. You can't judge a ripe mango from its skin color. Instead, use your smell and touch senses. Delicately squeezing a mango can help you determine if it's ripe or not, just like when you gently squeeze a peach to tell if it's ripe. Also, take a second to smell the fruit stem, which sometimes will have a fruity aroma when ripe. Once you've selected a ripe mango, wash it under water and pat it dry. Next, slice it in half by using a sharp knife to slice each side to the center seed. Then twist both sides in opposite directions. Remove the center stone. Then cut the pulp in each half into slices, being careful not to cut the mango skin (which contains a skin irritant). Then scoop out pulp slices and enjoy!

You can also freeze fresh ripe mangos. To freeze mangoes simply wash the mango with water then peel away the skin. Slice the pulp into slices or cubes. Lay the slices or cubes single layer on a tray, and place them in the freezer (set at zero degrees Fahrenheit or lower). Once they are thoroughly frozen, remove them from the tray and place them in an airtight freezer container.

They can also be frozen in a syrup pack. To freeze them in this method, prepare a 30 percent syrup pack by mixing one and three-quarters cup sugar into four cups of warm water. Stir the solution until it's thoroughly mixed then chill it. Place fresh mango slices in a canning jar. Pour the syrup pack over the fruit, covering it. Then seal the jar leaving adequate headspace.

(Headspace should be half-inch for a wide top pint jar, three-quarters of an inch for a narrow top pint jar, one inch for a narrow top quart jar, and one and a half inches for a wide top quart jar.) Place the jar in your freezer.

Fresh ripe mangos can also be dehydrated. Again, wash the mango then peel away the skin. Cut the mango pulp into slices preferably about a quarter-inch thick. Lay the slices on your electric dehydrator tray and set the temperature for 135 degrees to 140 degrees Fahrenheit. It can take anywhere from 9 to 12 hours or more to dry. To test for dryness, cut a couple of pieces in half. You should not be able to squeeze any moisture out of them, and there should be no visible moisture. And if you fold a piece over, it should not stick to itself. Once the fruit is dry, let it cool for an hour. Then package the fruit preferably in a freezer container or glass jar. Freezer bags can also be used but unfortunately they are not rodent proof. Store the container in a dark, dry, cool place for no more than a year.

Mangos are used in cakes, to make sorbet and pudding and in alcoholic beverages, in salads, and in cooked dishes. There are over 800 mango recipes listed on the Food Network's website. Mangos can also be canned and made into salsas, chutneys, and sauce. For information on how to can mangos, check out the National Center for Home Food Preservation's website at www.nchfp.uga.edu.

ADDITIONAL INFORMATION. Note that the mango tree leaves, sap, and fruit skin contain 3-pentadecyl catechol which usually causes an allergic reaction when you come in contact with it. The allergic reactions include an itching rash and swelling. So it's important to wear gloves and long sleeves when pruning mango trees to limit any skin contact with the tree leaves and sap. Likewise, when picking fruit, it's wise to wear gloves, and to wash any fruit to remove any sap on it.

Mango is the national fruit of India, Pakistan and the Philippines. India is the largest producer of mangos, producing approximately half of the global mango supply. They are followed by China, Thailand, Mexico, Indonesia, and Pakistan.

REFERENCES

Butterfield, Harry M. *A History of Subtropical Fruits and Nuts in California.* Berkeley, CA 1963.

Crane, Jonathan H., et al. "Mango Growing in the Florida Home Landscape." HS2. University of Florida, Institute of Food and Agricultural Sciences Extension. May 2017. https://edis. ifas.ufl.edu/pdffiles/MG/MG21600.pdf. Accessed 11/06/18.

D'Sa, Elaine M. "Mango—A Tropical Treat." National Center for Home Food Preservation, University of Georgia. August 2004. https://nchfp.uga.edu/publications/nchfp/factsheets/ the_mango.pdf. Accessed 11/12/18.

Folkhard, Richard. *Plant Lore, Legends and Lyrics: Embracing the Myths, Traditions, Superstitions, and Folk-Lore of the Plant Kingdom.* London: Sampson Low, Marston, Searle, and Rivington. 1884.

Gilman, Edward F., and Dennis G. Watson. "*Mangifera indica*: Mango." ENH563. University of Florida, Institute of Food and Agricultural Sciences Cooperative Extension. February 2013. https://edis.ifas.ufl.edu/pdffiles/ST/ST40400.pdf. Accessed 11/06/18.

Gilman, Edward F., and Dennis G. Watson. "*Mangifera indica* Mango." Fact Sheet ST-404. U.S. Forest Service, U.S. Department of Agriculture and Southern Group of State Foresters. October 1994. http://hort.ufl.edu/trees/MANINDA.pdf. Accessed 11/05/18.

"Growing Mango Trees: Information on Planting and Caring for a Mango Tree." Growing Know-How. www.gardeningknowhow.com/edible/fruits/mango/growing-mango-trees.htm. Accessed 11/06/18.

Hamilton, R.A., C.L. Chia, and D.O. Evans. "Mango Cultivars in Hawaii." Information Text Series 042. College of Tropical Agriculture and Human Resources, University of Hawaii, April 1992. https://scholarspace.manoa.hawaii.edu/bitstream/10125/42443/5/Mango%20Cultivars%20in%20Hawaii.pdf. Accessed 11/11/18.

Heen, Napua. "Mango on My Mind." Hawaii.com. https://www.hawaii.com/discover/mango/. Accessed 11/11/18.

"How to Grow Mango Trees from Seeds." Tropical Permaculture. www.tropicalpermaculture.com/growing-mangoes.html. Accessed 11/06/18.

Leitten, Rebecca Rose. "Plant Myths and Legends." Cornell University. http://bhort.bh.cornell.edu/conservatory/cpage3.html#Mango%20(Mangifera. Accessed 11/06/18.

"Mangifera Indica." Bioweb. University of Wisconsin System. http://bioweb.uwlax.edu/bio203/s2012/ruda_chel/classification.htm. Accessed 11/06/18.

"*Mangifera indica*." EcoCrop. U.N. Food and Agriculture Organization. http://ecocrop.fao.org/ecocrop/srv/en/cropView?id=1416. Accessed 11/05/18.

"*Mangifera indica*." Flora of the Hawaiian Islands. Botany Department, Smithsonian Institution. http://botany.si.edu/pacificislandbiodiversity/hawaiianflora/speciesdescr.cfm?genus=Mangifera&species=indica. Accessed 11/06/18.

"*Mangifera indica* L." Smithsonian Tropical Research Institute. https://biogeodb.stri.si.edu/biodiversity/species/30463/. Accessed 11/06/18.

"*Mangifera indica* L." USDA, NRCS. 2018. The PLANTS Database. National Plant Data Team, Greensboro, NC 27401-4901 USA. https://plants.sc.egov.usda.gov/core/profile?symbol=MAIN3. Accessed 12/15/18.

"Mango." Fruit Facts. California Rare Fruit Growers. http://www.crfg.org/pubs/ff/mango.html. Accessed 11/06/18.

"Mango." National Mango Board. https://www.mango.org/about-mangos/. Accessed 11/05/18.

"Mango General Crop Information." University of Hawaii Extension. www.extento.hawaii.edu/kbase/crop/crops/i_mango.htm. Accessed 11/06/18.

"Mango (*Magnifera indica*)." Department of Agriculture, Republic of South Africa, Pretoria. 2008. www.daff.gov.za/Daffweb3/Portals/0/Brochures%20and%20Production%20guidelines/Brochure%20Mango.pdf. Accessed 11/05/18.

Misra, R.K. "Gujarat's 'Walking' Tree Travels Through Time." *The South Asian Post*, June 23–29, 2011. p. 25.

Morton, Julia F. *Fruits of Warm Climates*. Eugene, OR: Wipf and Stock Publishers, 2003.

Ortho All About Citrus and Subtropical Fruits. Des Moines, IA: Meredith Books, 2008.

Orwa, C., et al. "Magnifera indica." 2009 Agroforestree Database: A Tree Reference and Selection Guide Version 4.0. www.worldagroforestry.org/treedb/AFTPDFS/Mangifera_indica. PDF. Accessed 11/03/18.

Pariona, Amber. "Top Mango Producing Countries in the World." World Atlas. April 9, 2018. www.worldatlas.com/articles/the-top-mango-producing-countries-in-the-world.html. Accessed 11/10/18.

"Philippine Myth on Mango Fruit." Philippines Insider. August 11, 2018. www.philippinesinsider.com/myths-folklore-superstition/philippine-myth-on-mango-fruits/. Accessed 11/06/18.

Popenoe, Wilson. *Manual of Tropical and Subtropical Fruits: Excluding the Banana, Coconut, Pineapple, Citrus Fruits, Olive and Fig*. NY: Macmillan Co., 1920.

Preserving Food: Drying Fruits and Vegetables. College of Family and Consumer Sciences, University of Georgia Cooperative Extension Services, July 2000.

Reed, Patricia Hamilton. "How to Grow Baby Mangoes Indoors. SFGate. https://homeguides.sfgate.com/grow-baby-mangoes-indoors-94188.html. Accessed 11/06/18.

"Seedless Mangos, Anyone?" *The South Asian Post*, July 31—August 16, 2014, p. SP11.
"SelecTree: Tree Detail MANGO *Mangifera indica*." Urban Forest Ecosystems Institute, California Polytechnic State University. https://selectree.calpoly.edu/tree-detail/mangifera-indica. Accessed 11/06/18.
Sengupta, Sushmita. "12 Interesting Mango Facts Even the Non Mango Lovers Would Enjoy!" Food. NDTV. April 13, 2018. https://food.ndtv.com/food-drinks/12-interesting-mango-facts-even-the-non-mango-lovers-would-enjoy-1837192. Accessed 11/6/18.
Singh, Nagendra K., et al. "Origin, Diversity and Genome Sequence of Mango (*Mangifera indica* L.)." *Indian Journal of History and Science*, vol. 51, no. 2.2, 2016, p. 355–368.

Nanking Cherry (*Prunus tomentosa* Thunb.)

Also known as a Manchu cherry, Chinese bush cherry, and Mongolian cherry, the Nanking Cherry generally bears a smooth, thin, shiny outer skin that ranges in color from pink to dark red (although a white skinned Nanking cherry cultivar exists). They have a soft juicy pulp, a single large hard stone in the center that encapsulates a seed. The cherry ranges from a third of an inch to three-quarters of an inch in diameter. The cherry is low in acid, and are generally tart and somewhat tangy. Its taste is sometimes compared to a sour cherry, however sweet cherries do exist. Their flavor is considered mild and most frequently eaten fresh.

NANKING CHERRY LORE. In China, cherry wood is believed to keep evil sprits at bay, so cherry branches are placed above doors on New Year's Day, and statues made of cherry wood are displayed at the entrances of homes to keep evil spirits out. In herbal folk medicine, Nanking cherries have been used as an anti-inflammatory to treat various inflammatory conditions ranging from arthritis to gout. They've also been used as an herbal supplement to prevent cancer.

PLANT HISTORY. The Nanking cherry originated in temperate eastern and central Asia. They were widely grown in the Russian Far East, China, Tibet and the Himalayas. In 1870 they were introduced into England. In 1882 specimens from China were collected by the U.S. Department of Agriculture and introduced into the United States. And subsequently in the early 20th century Frank Meyer of the USDA collected and sent 4200 seeds from China to the U.S. From the 1920s into the1930s there were efforts in America to create cultivars with sweeter larger pulp and smaller seeds. But those efforts were redirected to other fruits with greater market potential and a longer shelf life.

PLANT CHARACTERISTICS. Nanking cherry plants generally grow six to ten feet tall and equally wide, but can grow up to 15 feet tall and wide. A

Nanking cherry *(www.maxpixel.com)*

deciduous bush, its darker green serrated elliptic leaves have a somewhat fuzzy underside. Measuring from two to three inches long, and one to one and a half inches wide, the leaves are arranged alternately. On one year old and older growth, the plant bears pink buds that mature into fragrant white blossoms. The buds, measuring about three-quarters of an inch in diameter, appear in clusters. The plants are not self-fruitful so two or more plants are needed for cross-pollination, although self fertile plants supposedly exist.

The plant grows into an upright spreading, multi-stemmed dense twiggy bush. The plant's bark is shiny, reddish brown and exfoliating when mature. It has a medium depth root system and spread. Established plants are drought tolerant. The bush is attractive to bees, butterflies and birds. Plants are propagated by seed, by cuttings and by grafting. They generally begin fruiting after three or four years. Mature plants can produce up to 50 pounds of fruit per season. The plants can live up to 50 years.

▶ FAMILY: Rosaceae. GENUS: *Prunus* L. SPECIES: *Prunus tomentosa* Thunb.

SELECTING A PLANT. A number of different cultivars of Nanking cherries exist, but they may not be available in the United States. A sample of the cultivars include the 'Leucocarpa,' which produces white fruit, 'Pink Candles,' which produces small red cherries, the 'Orient,' which produces large orange red cherries, and 'Minnesota No. 41,' which produces large round fruit.

When selecting a bare root cherry plant from the nursery, check the bud union (i.e., the place where the cherry variety is grated to the rootstock). It should be straight. Avoid it if it is bent. Also avoid any plants that have dark colored oozing bark, which could indicate a potential canker caused by bacteria. Also check for any holes that could be caused by insects. And if possible, check to make sure the roots are not damaged.

How to Grow a Nanking Cherry Plant. Nanking cherry plants require full sun. They grow in most well draining soils but prefer loamy soils. They grow in soil with a pH of 5.0 to 7.5. Established plants are drought tolerant. Although some plants are self fruitful, you really do need at least two Nanking cherry plants for good pollination and fruit set if you are growing them for fruit, and simply not as ornamentals. Select a sunny location that can accommodate the mature size of your plants with six to seven feet of space between them. Most Nanking cherry plants are sold as bare root plants. When you receive or bring home your bare root cherry plants, you need to soak the plant roots in a bucket of water a minimum of two hours before planting them.

While the roots are soaking, you can dig a hole slightly deeper than the length of the plant roots, and twice the size of the root's diameter. When you are ready to plant your cherry plant, spread out the cherry plant roots and set the plant in the hole. Refill the hole with soil. The plant bud should be level with or slightly above the soil. Water the plant thoroughly. Water the plants regularly until they become established. The plants should be pruned annually to encourage new growth and to keep them from becoming too dense. Prune away a fourth to no more than a third of the branches each year.

How to Grow from Seed. Retrieve the seeds from fully ripe fruit and completely remove any pulp from the seed. Wash the seeds and let them dry in a well ventilated room for a few days. For germination, the seeds need to be cold stratified for 100 days. This can be done by keeping the seeds in consistently moist vermiculite or sphagnum peat moss in the refrigerator (32 to 45 degrees Fahrenheit). In the spring, after the cold stratification is complete, you can plant the seeds in pots outside. Some people also just plant the seeds outside in the fall and let the natural winter cold and rains stratify the seeds. Although this method is dependent upon the winter weather's providing regular moisture and the cold temperatures needed. Although they can begin fruiting at one year, generally they fruit in their third or fourth year.

▶ USDA Hardiness Zones: 2–7. Chilling Hours: 200–300. Water Requirements: Average.

How to Consume. Nanking cherries are frequently eaten fresh. They have a short shelf life of only a couple of days but you can keep Nanking cherries in a plastic bag in your refrigerator for up to a week. Nanking cherries can be used in baked goods, such as pies and breads. Fresh cherries can be dehydrated, but they need to be pre-treated first. For information on pre-treating and dehydrating cherries at home, check out the National Center for Home Food Preservation's website at https://nchfp.uga.edu/ for instructions.

Nanking cherries can also be canned and used to make jam and jellies. Instructions for preserving the cherries in these methods can also be found on the National Center for Home Food Preservation's website. The cherries can also be made into Nanking cherry wine. (There are instructions for making Nanking Cherry Wine on the Alaska Pioneer Fruit Growers Association website at https://www.apfga.org/featured-fruit-nanking-cherry/.)

ADDITIONAL INFORMATION. Nanking cherries contain lots of vitamins A and C, as well as calcium and iron.

REFERENCES

"The Chinese Bush Cherry—*Prunus Tomentosa.*" *Arnoldia: A Continuation of the Bulletin of Popular Information of the Arnold Arboretum, Harvard University*, vol. 24, no. 9, Sept. 18, 1964, p. 81–86. http://arnoldia.arboretum.harvard.edu/pdf/articles/1964-24—the-chinese-bush-cherry-prunus-tomentosa.pdf. Accessed 11/28/18.

Cumo, Christopher. *Encyclopedia of Cultivated Plants: From Acacia to Zinnia.* Santa Barbara, CA: ABC-CLIO. 2013.

"Featured Fruit—Nanking cherry." Alaska Pioneer Fruit Growers Association. www.apfga.org/featured-fruit-nanking-cherry/. Accessed 11/28/18.

Hamilton, Elizabeth, Tiffany Maughan, and Brent Black. "Nanking Cherry in the Garden." Horticulture/Fruit/2016-03pr. Utah State University Extension. August 2016. https://digitalcommons.usu.edu/cgi/viewcontent.cgi?article=2587&context=extension_curall. Accessed 11/28/18.

Hayman, Vicki, and Chris Hilgert. "No Lie! You Can Grow Cherries in Wyoming, Just Be Selective." *Barnyards and Backyards*, Summer 2016, p. 12–14. www.uwyo.edu/barnbackyard/_files/documents/magazine/2016/summer/cherries0716.pdf. Accessed 11/28/18.

Hill, Lewis, and Leonard Perry. *The Fruit Gardener's Bible.* North Adams, MA: Storey Publishing, 2011.

"Nanking Cherries." Bush Foundation Community Innovation Grant: Changing the Approach to Regulation of Local Food Systems in Minnesota http://misadocuments.info/nanking_cherries.pdf. Accessed 11/28/18.

"Nanking Cherries." Specialty Produce. www.specialtyproduce.com/produce/Nanking_Cherries_14833.php. Accessed 11/28/18.

"Nanking Cherry." Morton Arboretum. www.mortonarb.org/trees-plants/tree-plant-descriptions/nanking-cherry. Accessed 11/28/18.

"Nanking Cherry." North Dakota State University. www.ag.ndsu.edu/trees/handbook/th-3-11.pdf. Accessed 11/28/18.

"Nanking Cherry *Prunus tomentosa.*" Becker Soil and Water Conservation District, Becker County, Minnesota. www.co.becker.mn.us/dept/soil_water/PDFs/trees/Nanking%20Cherry.pdf. Accessed 11/28/18.

Preserving Food: Drying Fruits and Vegetables. College of Family and Consumer Sciences, University of Georgia Cooperative Extension Services, July 2000.

"Prunus Cultivar: Tomentosa." Foundation Plant Services, University of California, Davis. http://fps.ucdavis.edu/treedetails.cfm?v=1537. Accessed 11/28/18.

"*Prunus tomentosa* Thunb." USDA, NRCS. 2018. The PLANTS Database. National Plant Data

Team, Greensboro, NC 27401-4901. https://plants.usda.gov/core/profile?symbol=prto80. Accessed 11/28/18.

Reich, Lee. *Uncommon Fruits for Every Garden.* Portland: Timber Press, 2004.

"Species: *Prunus tomentosa.*" Woody Plants Database, Urban Horticulture Institute, School of Integrative Plant Science, Cornell University. http://woodyplants.cals.cornell.edu/plant/print/196. Accessed 12/03/18.

Stremple, Barbara Ferguson, editor. *All About Growing Fruits, Berries and Nuts.* San Ramon, CA: Ortho Books. 1987.

Noni (*Morinda citrifolia* L.)

Known by many names, including the Indian mulberry and cheese fruit, the noni fruit is oval shaped, measuring from about two to seven inches long with a lumpy texture and skin. Its size is often compared to that of a russet potato. Its rind is green when immature, and turns yellowish white when mature. The fruit pulp is off-white and somewhat gelatinous and has many small dark brown seeds surrounding the center. Unripe noni fruit is bitter. When it's ripe it emits a foul odor. And if you can get over its putrid odor, it reportedly has a bitter sharp flavor. Some people say fresh ripe noni is similar in taste to bleu cheese.

NONI LORE. Noni tree parts and fruit have a long history of being used in folk medicine in Polynesia and Asia. A liquid made from the tree roots has been used to treat urinary disorders, stiffness and skin diseases. A poultice made from the leaves is used to treat headaches. The juice is used to treat a wide variety of conditions from mouth ulcers and high blood pressure to diabetes and tuberculosis and leprosy. The fruit itself has been used to treat a wider range of maladies ranging from asthma to cancer. Although some preliminary studies have shown noni fruit may have anticancer, anti-inflammatory and analgesic properties, more studies need to be conducted. Although noni has some antioxidant qualities, for some people, consuming noni can pose health risks (see the Additional Information below).

TREE HISTORY. The noni is believed to have originated in Southeastern Asia and dates back to ancient times. It was first mentioned in Chinese literature during the Han dynasty (206 BCE–23 CE). Because noni seeds have air chambers and float, the seeds were spread by ocean currents and by early travelers to Australia and Polynesia and Seychelles. It was introduced into Hawaii when a Polynesian chief, Hawaii Loa, transported it in his canoe with him as he journeyed to a new island chain later named after him. It was regarded as a "canoe plant," i.e., one of the plants ancient Hawaiian's took with them on their canoes during their migrations.

In countries where it is currently grown it is often considered a "famine"

Noni *(Michael Hermann, www.CropsfortheFuture.org/Wikimedia Commons)*

food. Because of its foul odor and strong taste, it is eaten when little else is available for sustenance. It's regarded as an emergency food supply.

TREE CHARACTERISTICS. Noni trees are small tropical evergreen trees or shrubs that grow from 10 to 20 feet tall, with a canopy 8 to 15 feet wide. It has a straight trunk and grayish or yellowish brown bark. It bears large thick glossy dark green elliptic shaped leaves that are deep veined and measure from six to 20 inches long. It bears small white blossom clusters. The blossoms are bisexual, fragrant, and are pollinated by insects and to a smaller extent, the wind.

The tree has a deep taproot and an aggressive extensive root system. It's considered an invasive species in a number of countries, including Costa Rica, the Dominican Republic and Cuba. Mature noni trees are drought tolerant. The tree generally lives for a minimum of 25 years. The trees are susceptible to root knot nematodes.

The noni tree is propagated by seed, and by stem cuttings. Natural seed germination can take six months to a year or more. Stem cuttings usually root in one to two months. Trees grown from seed usually begin to blossom and fruit in their third year. There are currently no known cultivars of *Morinda citrifolia*.

▶ FAMILY: Rubiaceae. GENUS: *Morinda* L. SPECIES: *Morinda citrifolia* L.

SELECTING A TREE. When selecting a tree from the nursery, check the tree's overall health. Look for active growth, such as new or young leaves. Make sure its leaves are free from insects or signs of insect damage. Check the entire tree to make sure it is free from injury, such as broken branches, cuts in the bark, etc. If possible, lift the tree out of the pot and check its roots.

HOW TO GROW A NONI TREE. Noni trees can grow in a wide range of soils, including acidic and alkaline soils, but grows best in well draining soil. It prefers a soil pH of 5–6.5, but will grow in soil with a pH from 4.3. to 7. It prefers full sun and an annual daytime temperature between 75 and 86 degrees Fahrenheit. They do not grow well in areas with strong winds.

Select a sunny location or one with partial shade, shielded from the wind, that can accommodate the size of a mature tree. Dig a hole at least twice the diameter of the pot the tree came in, about as deep as the container. Place the tree in the hole, loosening or cutting away any girdled roots. The tree should be planted so that the top of the root ball is about one to two inches higher than the soil level. Refill the hole with the original soil. Water the newly planted tree thoroughly. Young trees benefit from regular watering when the soil is dry. But be careful not to over water the tree, since it's susceptible to root rot and root knot nematodes. After their first year of fruiting, noni trees may be pruned back. Pruned trees will become bushy. For mature trees, you may want to prune the vertical branches to facilitate fruiting.

HOW TO GROW FROM SEED. Remove the seeds from a fresh ripe, soft noni fruit. Remove all the pulp from around the seeds. (Using a colander to gather the cleaned washed seeds helps.) If allowed to germinate naturally, noni seeds can take up to a year to germinate, which is why seed scarification is recommended. There are many methods of seed scarification to choose from. Three popular methods of seed scarification are to nick it with a knife, rub it between sandpaper, or freeze it overnight then let it soak in hot water for several hours. Place the seeds in a moist planting mix. The seeds need a consistently hot (90–100 degrees Fahrenheit) moist environment for germination. The seeds should germinate in three to nine weeks. The seedling should be grown for at least nine months before being planted into the ground.

▶ USDA HARDINESS ZONES: 10–12. CHILLING HOURS: 0. WATER REQUIREMENTS: Prefers 60–118 inches rainfall annually.

HOW TO CONSUME. Ripe noni fruit can be eaten fresh, but most people consider it an acquired taste. Eating it with salt is supposed to make it more palatable. Fresh noni fruit are often juiced. Since the juice still retains its odor, it is frequently mixed with other fruit juices to improve the taste and mask the odor. (See warning below.)

Fresh ripe noni fruit is often processed and dried into a fruit leather. The taste does not improve; however, you can add a two-inch square of the fruit leather to your favorite cup of tea. (See warning below.) Noni fruit can also be cooked and added to dishes such as curries.

ADDITIONAL INFORMATION. WARNING: Drinking noni juice and tea regularly could potentially cause liver problems. There have been accounts of noni leading to acute liver injury. According to the National Institute of Health's LIVERTOX database, "Hepatotoxicity attributed to noni juice has occasionally been severe and led to acute liver failure, which in at least one case necessitated emergency liver transplantation. Rechallenge studies have not been reported."

According to the WebMd website, noni interacts with the following types of medications, so you should consult with your doctor before consuming noni if you are taking high blood pressure medication, Warfarin, water pills, and other medications that may also harm your liver. The root bark of the noni tree has also been used as a dye. At one time noni trees were planted specifically for their use in textile dyeing.

REFERENCES

"Datasheet: Morinda Citrifolia (Indian Mulberry)." CABI Invasive Species Compendium, CAB International. www.cabi.org/isc/datasheet/34854. Accessed 11/24/18.

Fern, Ken. "Morinda Citrifolia." Tropical Plants Database. http://tropical.theferns.info/ viewtropical.php?id=Morinda+citrifolia. Accessed 11/24/18.

"How Noni Fruit First Arrived to the Hawaiian Islands." Tru Noni. www.trunoni.com/12_ Noni_juice_history.htm. Accessed 11/24/18.

"How to Eat Noni Fruit." The Survival Gardener. March 29, 2017. www.thesurvivalgardener. com/eat-noni-fruit/. Accessed 11/24/18.

McClatchey, Will. "From Polynesian Healers to Health Food Stores: Changing Perspectives of *Morinda citrifolia* (Rubiaceae)." *Integrative Cancer Therapies*. Vol. 1, no. 2, 2002, p. 110–120. www.ctahr.hawaii.edu/noni/Downloads/MorindaCitrifolia.pdf. Accessed 11/24/18.

"*Morinda citrifolia*." Durable.org. http://www.doc-developpement-durable.org/file/Arbres-Fruitiers/FICHES_ARBRES/Morinda%20citrifolia-noni/Morinda%20citrifolia_noni.pdf. Accessed 11/24/18.

"*Morinda citrifolia*." Missouri Botanical Garden. www.missouribotanicalgarden.org/Plant Finder/PlantFinderDetails.aspx?taxonid=286670&isprofile=0&. Accessed 11/24/18.

"*Morinda citrifolia* L." Plants for a Future. https://pfaf.org/user/Plant.aspx?LatinName= Morinda+citrifolia. Accessed 11/24/18.

"*Morinda citrifolia* L." USDA, NRCS. 2018. The PLANTS Database. National Plant Data Team, Greensboro, NC 27401-4901 USA. https://plants.sc.egov.usda.gov/core/profile? symbol=MOCI3. Accessed 11/24/18.

Nelson, Scot C. "Noni Cultivation and Production in Hawai'i." From: *Proceedings of the 2002 Hawai'i Noni Conference*, S.C. Nelson (ed.), University of Hawaii at Manoa, College of Tropical Agriculture and Human Resources, 2003.

Nelson, Scot C. *Noni Cultivation in Hawaii*. F&N-4. University of Hawaii at Manoa, College of Tropical Agriculture and Human Resources, Cooperative Extension. March 2001. www. ctahr.hawaii.edu/oc/freepubs/pdf/F_N-4.pdf. Accessed 11/24/18.

Nelson, Scot C., and Craig R. Elevitch. *Noni: The Complete Guide for Growers and Consumers.* Holualoa, HI: Permanent Agriculture Resources, 2006. www.ctahr.hawaii.edu/UHMG/ downloads/2006-Noni-The-Complete-Guide-Nelson-Elevitch.pdf. Accessed 11/27/18.

"Noni." Canoe Plants of Ancient Hawai'i. www.canoeplants.com/noni.html. Accessed 11/24/18.

"Noni." LIVERTOX Database. National Institute of Health. https://livertox.nih.gov/Noni.htm. Accessed 11/27/18.

"Noni." WebMD. www.webmd.com/vitamins/ai/ingredientmono-758/noni. Accessed 11/24/18.

"Noni Fruit." Specialty Produce. www.specialtyproduce.com/produce/Noni_Fruit_12269.php. Accessed 11/25/18.

The Noni Website. University of Hawaii at Manoa, College of Tropical Agriculture and Human Resources, Cooperative Extension. www.ctahr.hawaii.edu/noni/. Accessed 11/24/18.

Orwa, C., et al. "Morinda citrifolia." 2009 Agroforestree Database: A Tree Reference and Selection Guide Version 4.0. www.worldagroforestry.org/treedb/AFTPDFS/Morinda_ citrifolia.PDF. Accessed 11/24/18.

Phalsa (*Grewia asiatica* L.)

Also known as a Falsa or Sherbert Berry, this small berry usually measures from one to two centimeters in size. The berry turns from green to a deep red or purple, almost black when ripe, similar in appearance to a blueberry. The berry's pulp is greenish white near the skin, turning purplish red starting from the seed and ultimately turning more purplish as it ripens. Each berry generally contains a single seed, with the larger berries usually having two seeds. Somewhat astringent, the berry has a short shelf life of only a few days when ripe. Its taste is described as pleasantly acidic.

PHALSA LORE. According to ancient medical treatises in India, phalsa berries can cure a variety of problems ranging from inflammation, to heart and blood disorders and fever. The berries are also said to be an aphrodisiac and can be made into a cooling beverage. The plant leaves are believed to have anticancer and antimicrobial properties. And the bark is made into an ointment that supposedly can cure urinary problems.

PLANT HISTORY. Native to India and Southern Asia from Pakistan to Cambodia, *Grewia asiatica* was later introduced in the early 1900s into the East Indies and the Philippines where it has become naturalized. It was also introduced into Australia where it has become naturalized in northern Australia. Later in the 1900s the plant made its way to the United States, initially both at the Agriculture Experiment Station in Puerto Rico and the University of Florida.

PLANT CHARACTERISTICS. Phalsa is a deciduous large shrub or small tree with mature plants growing from six to approximately 15 feet tall. The

Phalsa *(Miansari66/Wikimedia Commons)*

plant has long skinny droopy branches, and large alternate widely spaced thick green ovate shaped leaves measuring up to eight inches long and six inches wide. The underside of the leaves are covered with tiny hairs. The bark is grayish brown or grayish white. The phalsa plant bears small clusters of blossoms. There are usually three to five yellow blossoms per cluster. Each blossom has five petals and five sepals, and measures about two cm in diameter. The resulting berries are ready to be harvested approximately 45 to 55 days after blossoming. The plant is self fertile.

Phalsa plants are primarily propagated from seed. Propagation by air layering treated with a growth hormone is reported to have an 85 percent success rate, ground layering only having a 50 percent success rate. Propagation from semi-hardwood cuttings has the lowest reported success rate, of only 20 percent. The plants generally begin fruiting after two to three years. Mature plants produce approximately six and a half to 11 pounds of berries annually.

▶ FAMILY: Tiliaceae. GENUS: *Grewia* L. SPECIES: *Grewia asiatica* L.

SELECTING A PLANT. The tall plants growing in the wild produce very acidic berries. The low growing shrub plants grow berries that are less acidic.

Phalsa blossoms *(Asit K. Ghosh Thaumaturgist/Wikimedia Commons)*

There are no known cultivars of the plant. Phalsa plants are commonly prop-agated by seed, but only begin bearing fruit in their second or third year. Selecting a two or three year old plant will provide phalsa berries more quickly.

When selecting a plant from the nursery, check the plant's overall health. Look for active growth, such as new or young leaves. Make sure its leaves are not yellow or discolored, and are free from insects or signs of insect damage. Check the entire plant to make sure it is free from injury, such as broken branches, cuts in the trunk, etc. Gently lift the tree out of its pot to check the roots. Overgrown plants that have seriously girdled roots will likely develop a poor root structure so they should be avoided.

GROWING PHALSA. As a subtropical plant, it can withstand only a light frost. The plant should be grown in full sun. The plant grows in most soil types provided there's adequate drainage, but grows best in loamy soils. It prefers soil with a pH of 6.0 to 7.5, but will tolerate soil pH in the range of 5.5 to 8.

Select a sunny site that can easily accommodate the size of a mature plant. Dig a hole as deep as the container the plant came in, and twice the width. You want to plant the plant about the same depth as the container you purchased it in. When you place the plant in the hole, the top of the root ball should sit about one inch above the soil. Refill the hole with the original soil. Water thoroughly. The plant should be watered regularly to help it become established. If you're planting more than one plant, space them several feet apart from one another to accommodate the size of mature plants.

Increasing the phosphorous level in the soil is said to increase the sugar content of the berries. Because of their short shelf life, the berries should only be harvested when fully ripe. The plant requires annual pruning since it bears berries on the current year's growth. Phalsa can also be successfully grown indoors in a large container as long as it is placed in an area with abundant sunlight. If you grow it in a planter, make sure the planter is at least two feet in diameter and in depth to allow for root growth.

HOW TO GROW FROM SEED. To grow from seed, place the seeds in fertile well draining soil in a planter. Place the planter in the sun and water daily until plants begin to sprout. Seeds generally germinate in approximately 15 to 20 days. If you plan to grow phalsa from seed, note that seeds begin to lose their viability after 90 days. Take this into consideration if you plan to purchase phalsa seeds.

▶ USDA HARDINESS ZONES: 9–11. CHILLING HOURS: 0. WATER REQUIREMENTS: Requires 50–160 inches of rainfall annually, with 90–115 inches being optimal.

HOW TO CONSUME. When ripe, the berries can be eaten fresh as a dessert. They can also be juiced and enjoyed as a beverage. When used as a beverage, sugar is often added to sweeten the taste.

There are two common methods of making a refreshing phalsa beverage. The first method involves bringing a mixture of sugar and water to boil for about five to ten minutes. Remove the mixture from the stove. Add in washed phalsa berries and let them soak in the mixture overnight. The next morning strain out the berries and serve the beverage chilled. The second method is to wash the berries, and place them in the blender. Add water and sugar (usually a ratio of one to three of sugar to water, to a ratio of two parts sugar to three parts water is recommended.) Blend all the ingredients together. Strain the mixture then chill it or add ice. The berries are used commercially in soft drinks.

ADDITIONAL INFORMATION. The bark of the phalsa plant is used to make billiard cues, bows, golf club shanks, tool handles, and even rope. The phalsa plant is considered an environmental weed in Queensland and the Northern Territory of Australia, the Philippines, and Réunion Island (France).

REFERENCES

"AESA based IPM Phalsa." Department of Agriculture and Cooperation, Ministry of Agriculture and Farmers Welfare, Government of India. 2015. https://farmer.gov.in/imagedefault/ipm/phalsa.pdf. Accessed 12/09/18.

Aziz, Muhammad Maaz, et al. "Effect of Different Pruning Intensities and Times on Fruit Yield and Quality of Phalsa (*Grewia Asiatica* L.) *Journal of Agricultural Research*, vol. 56, no. 2, 2018, p. 107–111.

Cunningham, Sally. *Asian Vegetables: A Guide to Growing Fruit, Vegetables and Spices from the Indian Subcontinent*. Bath (UK): Eco-Logic Books. 2009.

Fern, Ken. "*Grewia asiatica*." Tropical Plants Database. http://tropical.theferns.info/view tropical.php?id=Grewia+asiatica. Accessed 12/09/18.

"*Grewia asiatica*." EcoCrop. United Nations Food and Agriculture Organization. http://ecocrop.fao.org/ecocrop/srv/en/cropView?id=6520. Accessed 02/14/18.

"*Grewia asiatica* L." USDA, NRCS. 2018. The PLANTS Database. National Plant Data Team, Greensboro, NC 27401-4901. http://plants.usda.gov/core/profile?symbol=GRAS2. Accessed 02/14/16.

"*Grewia asiatica* L." Weeds of Australia, Biosecurity Queensland Edition, Queensland Government, 2016. https://keyserver.lucidcentral.org/weeds/data/media/Html/grewia_asiatica.htm. Accessed 02/14/16.

"*Grewia asiatica* L., Malvaceae." Pacific Island Ecosystems at Risk. www.hear.org/pier/species/grewia_asiatica.htm. Accessed 02/14/16.

"*Grewia asiatica* 'Sherbet Berry.'" Agristarts. www.agristarts.com/index.cfm/fuseaction/plants.plantDetail/plant_ID/312/index.htm. Accessed 02/14/16.

"Growing Phalsa." Grow Plants. www.growplants.org/growing/falsa. Accessed 02/14/16.

Morton, Julia F. *Fruits of Warm Climates*. Eugene, OR: Wipf and Stock Publishers, 2003.

Orwa, C., et al. "*Grewia asiatica* L." 2009 Agroforestree Database: A Tree Reference and Selection Guide version 4.0. www.worldagroforestry.org/treedb/AFTPDFS/Grewia_asiatica.PDF. Accessed 05/02/18.

"Phalsa." Fruitipedia. www.fruitipedia.com/phalsa%20Grewia%20asiatica.htm. Accessed 02/14/16.

Singh, Krishan Kumar, and Shiv Pratap Sing. "Cultivation and Utilization in Phalsa (*Grewia asiatica* L.) under Garhwal Himalayas region." *Journal of Medicinal Plant Studies*, vol. 6, no. 1, 2018, p. 254–256.

"Taxon: *Grewia asiatica* L." U.S. National Plant Germplasm System. https://npgsweb.ars-grin.gov/gringlobal/taxonomydetail.aspx?id=18013. Accessed 07/11/16.

Yadav, A.K. "Phalsa: A Potential New Small Fruit for Georgia," p. 348–352. In: J. Janick (ed.), *Perspectives on New Crops and New Uses*. Alexandria, VA: ASHS Press, 1999. www.hort.purdue.edu/newcrop/proceedings1999/v4-348.html. Accessed 06/26/16.

Zia-Ul-Haq, Muhammad, et al. "*Grewia asiatica* L., a Food Plant with Multiple Uses." *Molecules*, vol. 18, 2013, p. 2663–2682. www.researchgate.net/publication/235755914_Grewia_asiatica_L_a_Food_Plant_with_Multiple_Uses. Accessed 05/02/18.

Pomegranate (*Punica granatum* L.)

Pomegranates are rounded fruit with a protruding calyx at the bottom of the fruit. Measuring from two and a half to five inches wide, it has a hard leathery rind. The rind turns from shades of pink to a deep purplish red as it matures. (Although less common, white rind and white pulp varieties do exist.) Inside the fruit are various compartments separated by bitter white spongy membranes. Each compartment is filled with small seeds each covered with a red translucent juicy pulp. The pulp is slightly tart and acidic.

POMEGRANATE LORE. Because pomegranates date back to prehistoric times, it is embedded in the lore of many nations and religions. For example, the origin of the pomegranate tree is linked to Bacchus, the Roman god of agriculture, wine and fertility. According to myth, a priest foretold that a nameless nymph loved by Bacchus would wear a crown. Hearing the priest's prophecy, she assumed she would become a queen. But the nymph, having lost her virgin honor, was transformed into a pomegranate tree by Bacchus. He twisted the calyx of the pomegranate blossom into the form of a crown, to symbolize the crown the priest foretold she would wear.

But according to second century Greco-Roman poet Oppian, the origin of the pomegranate tree is attributed to a different legend. In this legend, a man who lost his first wife fell in love with his daughter Side. To escape her father's cruel persecution, she took her own life. The gods, feeling sympathy for Side, transformed her body into a pomegranate tree, and punished her father by turning him into a sparrow hawk that would forever shun the tree.

The pomegranate is prominent in Greek mythology. In the story Persephone, the daughter of Zeus and Demeter, is out picking flowers one day when, Hades, the god of the underworld dragged her down to the underworld to be with him. While she was in the underworld, Persephone ate some pome-

Pomegranate *(inaquim/Wikimedia Commons)*

Pomegranate plant *(Tubifex/Wikimedia Commons)*

granate seeds, symbolic of binding herself to Hades. Her mother, Demeter, the harvest goddess, missed her daughter so much that she prevented any food from growing. That's when Zeus intervened and ruled that Persephone would live three months in the underworld with Hades, and the remaining nine months on earth with Demeter. When Persephone lived above the earth, the harvest would once again grow. But during the three months Persephone was back in the underworld, during winter, the harvest would not grow.

In Buddhism, Hariti was a cannibalistic ogress who had several hundred children she loved. But to feed them, she abducted other children, and killed them to feed her children. According to lore, Buddha cured her of this habit by feeding her a pomegranate. According to Jewish lore, a pomegranate fruit has 613 seeds, one for each of the 613 commandments in the Torah. These are just a few of the many myths and legends associated with the pomegranate.

TREE HISTORY. Most experts agree that the pomegranate tree originated in Persia (now Iran) and adjacent areas. Wilson Popenoe writes in his book, "The cultivation of the pomegranate, which began in prehistoric times, was extended, before the Christian era, westward toward the Mediterranean and eastward into China." From Iran, the pomegranate was introduced into India around the first century CE. It was recorded in Bermuda in 1621. Spanish settlers introduced the tree into Arizona in the 16th century and into California much later in 1769. Today pomegranate trees are widely grown in India, Iran, China, Turkey, Spain, the United States, Afghanistan, Pakistan, Uzbekistan, Israel, the Middle East, dry regions of Southeast Asia, peninsular Malaysia, the East Indies and tropical Africa. India and Iran are the top pomegranate producers.

PLANT CHARACTERISTICS. Grown naturally, the pomegranate is a bush or shrub growing generally 15 to 20 feet high, but can grow up to 30 feet tall. But it can be trained to grow as a tree. Pomegranates grown commercially are usually trained as trees. Dwarf varieties exist, but are usually grown as ornamentals rather than for fruit production. The plant suckers from the base and the plant becomes very dense with multiple stems if they are not removed. The bark is dark gray. It's a deciduous or semi-deciduous plant depending upon the region. The leaves are lance-shaped or oblong and measuring an average of three inches long, bright green and somewhat glossy.

The plant bears striking orangey red blossoms that are borne on the tips of new branches either singly or in small clusters of up to five. Each blossom measures one and a quarter to two and a third inches wide, has five to seven lance-shaped petals with a tubular calyx. The plant is self pollinated but cross pollination increase the amount of fruit produced. Pollination is primarily done by insects, more than the wind.

Pomegranate vigor reportedly declines after about 15 years. But the plant has a long lifespan, with some plants in Europe recorded as living over 200 years. Pomegranates are often grown as a landscape plant. Pomegranate plants are propagated by seed, by hardwood cuttings, and to a lesser extent, air layered from suckers. Although plants can begin bearing fruit a year after planting, it's more common for them to begin bearing at three to four years. Mature trees can produce 250 or more fruit annually.

▶ FAMILY: Punicaceae. GENUS: *Punica* L. SPECIES: *Punica granatum* L.

SELECTING A TREE. Due to their long history and widespread growth, there are hundreds of pomegranate cultivars, but fewer than two dozen are available in the United States. Popular cultivars include the following: 'Wonderful'—produces large dark red skinned fruit. • 'Grenada'—produces a medium pink skinned fruit. • 'Ambrosia'—produces large pink skinned fruit. • 'Eversweet'—produces sweet red skinned fruit with soft seeds. • 'Sweet'— produces very sweet pink skinned fruit.

When selecting a tree from the nursery, check the plant's overall health. Look for active growth, such as new or young leaves. Make sure its leaves are not yellow or discolored, and free from insects or signs of insect damage. Check the entire tree to make sure it is free from injury, such as broken branches, cuts in the trunk, etc. And if possible, gently lift the tree out of the pot to check the roots. Overgrown plants that have seriously girdled roots will develop a poor root structure so they should be avoided.

How to Grow a Pomegranate Bush/Tree. Pomegranate plants produce their best fruit in regions with long hot summers and cool winters. They grow best in well drained soil, but can be grown in a wide range of soils from calcareous soil to acidic loam. The plant prefers a soil pH of 5.5 to 7.0. They require full sun for best fruit production. However they will grow in part shade, but will produce inferior fruit.

Select a sunny location that can accommodate the size of a mature pomegranate shrub/tree. Dig a hole at least twice the diameter of the pot the tree came in, about as deep as the container. Place the tree in the hole, loosening or cutting away any girdled roots. The tree should be planted so that the top of the root ball is about one to two inches higher than the soil level. Refill the hole with the original soil. Water the newly planted tree thoroughly. During dry weather the newly planted pomegranate should be watered every two to four weeks until it is established. Established pomegranate plants are drought tolerant, but should be watered for best fruit production. Young plants are also often fertilized with a balanced nitrogen fertilizer in the spring to help with fruit production and plant vigor.

If you prefer to train your pomegranate to grow as a tree, select one trunk and prune away any suckers on an ongoing basis. Allow the growth of three to five scaffold branches starting at about ten inches above the soil and spaced four to six inches apart. To grow as a shrub, allow four to six shoots to develop from the stem, keeping the plant balanced. Prune away any suckers and any other shoots to retain a vase shaped shrub. Ongoing pruning is only needed to remove any suckers and dead branches.

There are a variety of pomegranate pests. A good source for control of pomegranate pests is the University of California IPM Pest Management Guidelines for Pomegranates which can be accessed online at http://ipm. ucanr.edu/PDF/PMG/pmgpomegranate.pdf.

How to Grow from Seed. Pomegranates grow easily from seed, but plants grown from seed may not bear true to type. To grow a plant from seed, remove and wash away any pulp from the seeds. Then let the seeds dry for a few days to prevent them from rotting during the germination period. Plant the seeds, no deeper than quarter-inch, in a pot with moist potting soil. Place the pot in a warm sunny location, keeping the soil moist during the germination process.

How to Grow from Hardwood Cuttings. Pomegranates are most frequently grown from cuttings. To grow from a cutting, in the fall, take a cutting about a quarter- to half-inch in diameter and about 10 inches long from one year old wood or suckers. A rooting hormone, although not required, can be placed on the bottom of the cutting. Place the cutting in the soil vertically, leaving two to three inches of the top of the cutting above the soil.

▶ USDA Hardiness Zones: 8–10, although cold hardy cultivars do exist. Chilling Hours: 100–150. Water Requirements: Average.

How to Consume. Pomegranates can be eaten fresh. Be sure to select mature pomegranates to consume fresh. Mature pomegranates make a metallic sound when tapped. You also want to select heavy fruit, indicating juicy arils. When eating a fresh pomegranate, you may want to wear an apron since the seeds stain easily and the stains are hard to remove. To eat a fresh pomegranate, slice away the protruding calyx. Next carefully slice the fruit rind from top to bottom. Then gently pull the fruit open so as not to splash juice on unwanted surfaces. Now you can pull out the pulp covered seeds or scoop them out with a spoon and consume the seed pulp. The pomegranate arils are also often added to salads.

You can flash freeze the pomegranate seeds for future addition to salads. Simply lay out the arils single layer on a cookie pan, place the pan in your freezer (zero degrees Fahrenheit or lower), until they are frozen. Then place them in a freezer container and put it back into the freezer. For other frozen uses, you can also freeze them in a syrup pack. To freeze them in this method, prepare a 30 percent syrup pack by mixing one and three-quarters cup sugar into four cups of warm water to dissolve the sugar. (If you desire, you can increase the syrup sweetness by increasing the amount of sugar.) Stir the solution until it's thoroughly mixed, then chill it. Place fresh pomegranate arils in a canning jar. Pour the syrup pack over the arils, covering them. Then seal the jar leaving adequate headspace. (Headspace should be half-inch for a wide top pint jar, three-quarters of an inch for a narrow top pint jar, one inch for a narrow top quart jar, and one and a half inches for a wide top quart jar.) Place the sealed jar in your freezer set at zero degrees Fahrenheit or lower.

Fresh pomegranates can be kept for up to two months at 32 to 41 degrees Fahrenheit. They are also grown for their juice. To secure the juice for drinking, slice the pomegranate in half and use a citrus juicer to extract the juice. You can also dehydrate pomegranate seeds in an electric dehydrator. The resulting pomegranate "raisins" can be eaten as a snack. They can also be used in baked goods. Strained pomegranate juice can also be used to make pomegranate jelly.

ADDITIONAL INFORMATION. Pomegranates have antioxidants. Through the ages pomegranates have been claimed to cure a variety of illnesses. Various studies on the effectiveness of pomegranates in treating various conditions are ongoing.

REFERENCES

Castle, William S., James C. Baldwin, and Megh Singh. "Pomegranate in Florida for Commercial Enterprises and Homeowners." *Proceedings of the Florida State Horticultural Society*, no. 124, p. 33–40. 2011. www.journals.fcla.edu/fshs/article/download/84102/81735. Accessed 11/14/18.

Crites, Alice M., and Mary Wilson. "Pomegranate Fruit and Juice." Fact Sheet 05-36. University of Nevada, Reno, Cooperative Extension. 2005. www.unce.unr.edu/publications/files/hn/2005/fs0536.pdf. Accessed 11/14/18.

Folkhard, Richard. *Plant Lore, Legends and Lyrics: Embracing the Myths, Traditions, Superstitions, and Folk-Lore of the Plant Kingdom.* London: Sampson Low, Marston, Searle, and Rivington. 1884.

Heflbower, Rick, and Robert Morris. "Pomegranate, Fruit of the Desert." Utah State University Cooperative Extension. May 2013. https://digitalcommons.usu.edu/cgi/viewcontent.cgi?article=1322&context=extension_curall. Accessed 11/14/18.

Holland, D., K. Hatib, and I. Bar-Ya'akov. "Pomegranate: Botany, Horticulture, Breeding." *Horticultural Reviews*, vol. 35, 2009. https://ucanr.edu/sites/pomegranates/files/164442.pdf. Accessed 11/14/18.

Johnson, J.F. *Pomegranate Growing.* New South Wales Department of Agriculture. 1983. www.dpi.nsw.gov.au/__data/assets/pdf_file/0005/119543/pomegranate-growing.pdf. Accessed 11/14/18.

Lamborn, Alicia R. "Pomegranate." University of Florida, Institute of Food and Agricultural Sciences Extension Baker County. 2017. https://sfyl.ifas.ufl.edu/media/sfylifasufledu/baker/docs/pdf/horticulture/educator-resources/Pomegranate.pdf. Accessed 11/18/18.

MacLean, Dan, Karina Martino, and Harald Scherm. "Pomegranate Production." Circular 997. University of Georgia Cooperative Extension. March 2017. https://secure.caes.uga.edu/extension/publications/files/pdf/C%20997_5.PDF. Accessed 11/18/18.

Morton, Julia F. *Fruits of Warm Climates.* Eugene, OR: Wipf and Stock Publishers, 2003.

Ortho All About Citrus and Subtropical Fruits. Des Moines, IA: Meredith Books, 2008.

Orwa, C. "Punica granatum." 2009 Agroforestree Database: A Tree Reference and Selection Guide Version 4.0. www.worldagroforestry.org/treedb/AFTPDFS/Punica_granatum.PDF. Accessed 11/15/18.

"Pomegranate." California Rare Fruit Growers. www.crfg.org/pubs/ff/pomegranate.html. Accessed 11/14/18.

"Pomegranate." Fruit and Nut Research and Information, University of California, Davis. http://fruitandnuteducation.ucdavis.edu/fruitnutproduction/Pomegranate/. Accessed 11/14/18.

"Pomegranate." Medical University of South Carolina. http://academicdepartments.musc.edu/ohp/urban-farm/crop%20sheets/Pomegranate%20Fact%20Sheet%20with%20Recipe%20Nov%202016.pdf. Accessed 11/14/18.

"Pomegranate." University of Hawaii, College of Tropical Agriculture and Human Resources. www.ctahr.hawaii.edu/sustainag/extn_pub/fruitpubs/pomegranate.pdf. Accessed 07/25/18.

"Pomegranates." Urban Harvest. http://urbanharvest.org/documents/118591/8495086/Pomegranates+-+2018.pdf/0e086ba8-d6eb-4af0-a69e-304efc7d81ee. Accessed 11/14/18.

Popenoe, Wilson. *Manual of Tropical and Subtropical Fruits: Excluding the Banana, Coconut, Pineapple, Citrus Fruits, Olive and Fig.* NY: Macmillan Co., 1920.

"Punica granatum L." USDA, NRCS. 2018. The PLANTS Database. National Plant Data Team, Greensboro, NC 27401-4901 USA. https://plants.sc.egov.usda.gov/core/profile?symbol=PUGR2. Accessed 11/18/18.

Sheets, M.D., M.L. Du Bois, and J.G. Williamson. "The Pomegranate." HS 44. University of

Florida, Institute of Food and Agricultural Sciences Extension. March 2015. http://edis.ifas. ufl.edu/pdffiles/MG/MG05600.pdf. Accessed 11/18/18.

Stein, Larry, Jim Kamas, and Monte Nesbitt. "Pomegranates." Texas A&M University, AgriLife Extension. https://aggie-horticulture.tamu.edu/fruit-nut/files/2010/10/pomegranates.pdf. Accessed 11/14/18.

Stover, Ed, and Eric W. Mercure. "The Pomegranate: A New Look at the Fruit of Paradise." *HortScience*, vol. 42, no. 5, August 2007, p. 1088–1092. https://naldc.nal.usda.gov/download/ 7460/PDF. Accessed 11/14/18.

Tanner, Cory. "Pomegranate Factsheet." HGIC 1359. Clemson University Extension, Home and Garden Information Center. January 27, 2016. https://hgic.clemson.edu/factsheet/ pomegranate/. Accessed 11/15/18.

"UC IPM Pomegranate Pest Management Guidelines for Agriculture." Integrated Pest Management, Agriculture and Natural Resources, University of California. June 2017. http:// ipm.ucanr.edu/PDF/PMG/pmgpomegranate.pdf. Accessed 11/14/18.

Pomelo (*Citrus maxima* [Burm. f.] Merr.)

If you see what looks like a grapefruit on steroids in the supermarket, it's likely a pomelo. Pomelos generally tend to be two to three times the size of the common grapefruit. On average they weigh three pounds and measure six to eight inches in diameter. Known by several other names, including pumelo, pummelo, pompelmous, Chinese grapefruit, lucban, and at one time shaddock, pomelos can be round, somewhat pear shaped, or somewhat oblate. Thought of as the ancestor of the grapefruit, this citrus fruit has an extremely thick pith (the soft fibrous tissue underneath the rind). Because of the thick peel, pomelos can be stored for long periods of time. Pulp colors range from yellow, to pink, to white. The pulp generally tends to be sweeter than a grapefruit, and less acidic, but not as juicy. It has only a few large seeds. Approximately two dozen cultivars exist. Popular cultivars include 'Oroblanco,' the pink fleshed 'Chandler,' the 'Cocktail,' and the hybrid 'Melogold.'

POMELO FOLKLORE. In Asian culture, pomelos are considered a good luck fruit. In Mandarin, the pronunciation for "pomelo" and "blessing" are the same. During Chinese New Year, you'll often find pomelos and oranges on display in homes to bring in good fortune. In Asia you'll also encounter them in homes during the Mid-Autumn Festival (a.k.a. Moon Festival) as an offering for the Moon Goddess.

The pomelo has also been used in Asia in moving ceremonies. Pomelo leaves are soaked in water then scattered around a new home. The ceremony is supposed to drive away bad luck. In Honduras, pomelos play a role in an old wives tale. Babies who constantly cry are thought to be plagued by evil spirits. A single pomelo leaf is placed under the baby's mattress to dispel the evil spirits. Historically in Asia pomelos were used to treat coughs, fever and gastrointestinal problems. Currently there are claims that pomelos can cure asthma, lower high cholesterol, treat a variety of skin conditions, prevent

Pomelo *(terimakashi/pixaby.com)*

cancer, and cure headaches. But there's presently insufficient research to substantiate its effectiveness for treatment of any of these conditions.

TREE HISTORY. Experts believe the pomelo originated in either Southeast Asia (Thailand, Malaysia) or in Polynesia and the Malay Peninsula. It was introduced into China around 100 BCE. It's believed to have been introduced into the New World in the 17th century when Captain Shaddock of the East India Company who brought a number of citrus seeds, including the pomelo, to Barbados. (This is why the pomelo used to be called a shaddock, after Captain Shaddock).

The USDA started importing some trees from Asia starting in 1902 and sent bud wood to USDA facilities in Florida, California, and other selected citrus growing states. But that first group of imports and a series of subsequent imports yielded poor quality fruit. It wasn't until 1912 that a tree from Thailand was imported to the U.S. which ultimately provided bud wood, which was grafted onto rootstock that provided high quality pomelo fruit. Since then, many new cultivars have been developed in the United States. Cultivars like the 'Chandler,' 'Cocktail,' 'Valentine' and hybrid 'Melogold' were all developed at the Citrus Research Center at the University of California in Riverside.

TREE CHARACTERISTICS. Like other evergreen citrus trees, the pomelo tree develops very fragrant blossoms. In the United States they generally grow 15 to 25 feet tall. In perfect conditions in other countries they can grow up to 50 feet tall and proportionately wide. They tend to have a slightly crooked trunk. Dwarf varieties are available if space is an issue. Its dense branches and canopy make a pomelo tree a nice shade tree or a landscape tree. The trees can also be espaliered. Like other citrus trees, the leaves are oblate and somewhat glossy.

▶ FAMILY: Rutaceae. GENUS: *Citrus* L. SPECIES: *Citrus maxima* (Burm. f.) Merr.

SELECTING A TREE. Most nurseries that carry pomelo trees usually offer only a few cultivars. The 'Chandler' and the 'Oroblanco' appear to be the two most available cultivars. However, if you live in California, and if you know how to graft bud wood, you can order bud wood from 11 different pomelo varieties for a very small fee from the Citrus Clonal Protection Program (CCPP) at the University of California at Riverside. For more information on the CCPP and how to order bud wood, check out their website at www.ccpp.ucr.edu. Pomelo trees can also be grown from seed. The fruit from trees grown from seedlings tend to be smaller than the original fruit. It can also take trees grown from seed up to ten years to produce fruit. Pomelo trees can also be vegetatively propagated.

When selecting a tree from the nursery, check the tree's overall health. Make sure its leaves are not discolored, and free from insects or signs of insect damage. Check the entire tree to make sure it is free from injury, such as broken branches, cuts in the trunk, etc. Check to make sure the tree trunk is not sunburned. Also visually inspect the pot soil to make sure there is no fungi growing from the soil, indicating uneven or over watering that can affect the overall health of the trees. If possible, gently lift the tree from the pot to inspect the roots. Overgrown pot-bound trees with seriously girdled roots should be avoided since they are likely to develop a poor root structure.

GROWING POMELOS. Pomelos tolerate a wide range of soils providing there's adequate drainage. And like other citrus trees, they thrive in sunny warm climates provided they receive adequate water. Trees should be planted in full sun and regularly irrigated to promote root growth. If growing more than one in your yard, they should be spaced at least 10 to 15 feet apart to provide adequate space for canopy and root growth.

To plant a tree, dig a hole at least twice the diameter of the pot the tree came in, and at least as deep as the container. Place the tree in the hole, filling the hole with the original soil. The tree should be planted so that the top of the root ball is about one to two inches higher than the soil level/grade. Water thoroughly. The tree should be watered regularly until it becomes established. As a citrus tree, it will benefit from fertilization. Apply fertilizer (February—September) three times a year during the first two years, then twice a year after that. Pomelo trees only need pruning if you want to control the size of the tree. Also prune to remove any overcrowded branches. Pruning is best done in the spring after all fruit has been harvested.

Young citrus trees are sensitive to frost. If temperatures are expected to dip to 29 degrees Fahrenheit or lower, you need to protect the tree from frost. You should wrap the tree trunk and branches in insulation material like cardboard, fiberglass or a frost protection covers. Make sure the soil is moist and not dry, since damp soil retains and radiates more heat than dry soil. Bare soil also radiates more heat than soil covered with mulch or other ground covers. If your tree does sustain frost damage, don't prune away any dead branches until the spring. This allows time for the tree to recover in warmer weather. It also allows you to better identify the damaged branches to be removed.

HOW TO GROW FROM SEED. To grow a pomelo tree from seed, obtain seeds from a fresh pomelo and wash any pulp from them. Soak the seeds in water for a few hours. You want to plant the seeds in a pot with a rich medium that retains water, like soil mixed with perlite or peat moss. The general rule for planting any seed is to plant it twice as deep as the size of the seed. Once you plant the seed, if you have a green house, place it in the greenhouse. If not, cover the pot with plastic wrap to retain moisture and place the pot in a warm sunny place. Citrus seeds germinate best at soil temperatures around 80 degrees Fahrenheit. Don't allow the soil mix to dry out during the germination period, which can take up to eight weeks.

▶ USDA HARDINESS ZONES: 9–11. CHILLING HOURS: 0. WATER REQUIRE-MENTS: Average.

HOW TO CONSUME. Pomelos are most commonly eaten fresh. You'll frequently find them added to salads when in season, and as garnishes for main dishes. To eat a fresh pomelo, wash it under running water and dry it with a paper towel. Slice off the top of the fruit down to the flesh. Then carefully cut the rind and pith down from top to bottom of the fruit careful not to cut the pulp. Make four to six cuts of these cuts around the pomelo. Then peel the pith and rind away from the flesh. Pull apart each individual flesh segment and peel away the outer membrane, which is thick unlike that of an orange, to consume the pulp. Like other citrus fruit, the peel is often candied. If you've never candied citrus peel before, following is the simple method I use. There are many different methods and ingredients people use to candy citrus peel which you can find online if you want to experiment with them to find the method you like best.

Candied Pomelo Peels. Ingredients: An equal amount of peel, sugar and water. Wash the pomelo, rubbing it under running water. Dry it with a paper towel. Slice the peel into strips. Remove as much of the pith from the peel as possible. To remove the bitterness from the peel, put the peel slices in a pot of water and boil for 15 minutes. Drain the water and repeat the boiling for two more times. Now it's time to make the sugary syrup. Mix the sugar into the water and cook at medium low heat until the sugar melts. Put the peel strips in the pot and continue cooking until the water evaporates (this will take a couple of hours). Then carefully pour out the strips onto a rack to cool.

ADDITIONAL INFORMATION. WARNING: Pomelos contain many of the same substances found in grapefruits, including an enzyme that breaks down a variety of medications. Therefore you should consult with your physician if you are taking any of the following types of drugs which may interact with pomelos: statins, calcium channel blockers, other cardiovascular drugs, immunosupressants, sedatives, other psychiatric medications, and medications for erectile dysfunction. One cup of pomelo sections is approximately 72 calories. Pomelos are high in vitamin C.

REFERENCES

Barrett, Herb C. "ARS Releases Breeding Line of Pummelo, Exotic Citrus." *Agricultural Research*, vol. 42, no. 7, July 1994, p. 23.

"*CITRUS maxima.*" Learn2Grow. www.learn2grow.com/plants/citrus-maxima/. Accessed 01/28/16.

"*Citrus maxima* (Burm. f.) Merr.." USDA, NRCS. 2018. The PLANTS Database. National Plant Data Team, Greensboro, NC 27401-4901. https://plants.sc.egov.usda.gov/core/profile?symbol=CIMA5. Accessed 12/03/18.

"*Citrus maxima* (Burman) Merr.." California Department of Food and Agriculture. www.cdfa.ca.gov/plant/pe/AgCommID/page35.htm. Accessed 01/28/16.

"Crop Info and How-To Guide in Growing Pummelo." Cropsreview.com. www.cropsreview.com/pummelo.html. Accessed 11/14/2005.

David, Cynthia. "Fresh Bites: Pomelo." *Toronto Star*, January 2015.

Derksen, Joy. "Pomelo." *The Master Gardening Bench*, vol. 11, no. 2, February 2002, p. 1.

Franklin, Rosalind. *Baby Lore: Superstitions & Old Wives Tales from the World Over Related to Pregnancy, Birth & Babycare.* Diggory Press, 2005.

Geisel, Pamela M., and Carolyn L. Unruh. *Frost Protection for Citrus and Other Subtropicals.* ANR Publication 8100. University of California, Division of Agriculture and Natural Resources, 2003.

"Grapefruit, Drug Interactions." *Harvard Health Letter*, November 2010. www.health.harvard.edu/newsletter_article/grapefruit-drug-interactions. Accessed 06/02/18.

"Grapefruits and Pomelos." *Country Living*, vol. 20, no. 12, December 1997, p. 150.

Kelley, Jeanne. "Pomelos, Grapefruit's Sweeter and Mellower Relative, Have a Wealth of Flavor," *Los Angeles Times*, Feb. 12, 2016. www.latimes.com/food/la-fo-pomelo-20160213-story.html. Accessed 08/01/2017.

Lee, Thomas. "ASK THE DOCTOR. Does Pomelo Juice Affect Drugs the Same Way Grapefruit Juice Does?" *Harvard Heart Letter*, vol. 2, no. 3, November 2010, p. 7.

Lim, T. K. *Edible Medicinal and Non-Medicinal Plants, vol. 4.* New York, NY: Springer, 2012.

Morton, Julia F. *Fruits of Warm Climates.* Eugene, OR: Wipf and Stock Publishers, 2003.

"Moving Ceremony—China." USC Digital Folklore Archives. http://folklore.usc.edu/?p=2531. Accessed 01/26/16.

O'Connell, Neil V., et al. *Sample Costs to Establish a Pummelo Orchard and Produce Pummelos: Specialty Citrus.* CS-VS-02-3. San Joaquin Valley, CA: University of California Cooperative Extension, 2002.

Ortho All About Citrus & Subtropical Fruits. Des Moines: Meredith Books. 2008.

Orwa, C., et al. "*Citrus maxima.*" 2009 Agroforestree Database: A Tree Reference and Selection Guide Version 4.0. www.worldagroforestry.org/treedb/AFTPDFS/Citrus_maxima.PDF. Accessed 01/28/16.

Sterman, Nan. "Grow your own exotic citrus—pummelos, blood oranges and kaffir limes." *Los Angeles Times*, July 3, 2008. www.latimes.com/style/la-hm-citrus3-2008jul03-story.html. Accessed 01/28/16.

"Vitamins: Grapefruit." WebMd. www.webmd.com/vitamins/ai/ingredientmono-946/grapefruit. Accessed 06/02/18.

Pulasan (*Nephelium mutabile* Blume, *synonym Nephelium ramboutan-ake*)

Also known as a purasan, bulala, and the "hairless" rambutan, the pulasan is closely related to the rambutan and lychee. The pulasan is slightly larger than the rambutan and sweeter, although less juicy. The pulasan is egg shaped or oblong. On average it measures two to three inches wide. Upon first glance, it looks like a rambutan with a haircut. The skin is covered with short, thick, spines, opposed to the rambutan which has long hairy spines. With a red, or sometimes yellow skin depending upon the cultivar, the pulasan has a thick skin up to three-eighths of an inch thick. The interior flesh is translucent white. There's a single brown somewhat flattened almond shaped seed in the middle, although seedless cultivars do exist. Pulasan tastes like a sweeter version of a rambutan or lychee. The pulasan seed, unlike the rambutan seed, can be eaten raw. Its taste is like that of an almond.

PULASAN LORE. In the Philippines there is old weather lore that when the pulasan begins to ripen, the wet season begins. The farmers use the ripening of the pulasan as a sign it's time for them to plant their first crop of rice. According to folk medicine, a decoction of the pulasan roots supposedly gets rid of intestinal worms. And when used in a poultice, the leaves and roots are supposed to get rid of fevers.

TREE HISTORY. Originating in western Malaysia, the tree has not gained much popularity outside of Southeast Asia. It's also grown in the Philippines and India. It was introduced into Puerto Rico in 1926 and into Honduras in 1927. Occasionally the tree is grown in Costa Rica.

Pulasan *(Michael Hermann, www.CropsfortheFuture.org/Wikimedia Commons)*

TREE CHARACTERISTICS. Considered an ultra tropical evergreen tree, it grows up to 33 to 50 feet all, with a trunk measuring up to 18 inches in diameter. It has the same general form as the rambutan tree, only smaller. The tree has dark green glossy pinnate alternate leaves measuring seven to 18 inches long. It bears small greenish blossoms with no petals. The blossoms are borne either individually or in clusters. The blossoms are pollinated by bees and other insects. Trees grown from seed are either male or female. Male trees do not produce fruit. Only a few male trees are required to pollinate female trees. And trees grown from seed may not bear true to type. Trees grown from seed usually take about seven years before bearing fruit.

Budding and grafting are the preferred methods of propagation. Trees propagated by these two methods produce predominantly bisexual blossoms and a small number of male blossoms. So male trees are often planted in orchards to make sure there's available pollen for female blossoms. Trees propagated in these methods tend to begin fruiting in as little as three to five years.

▶ FAMILY: Sapindaceae. GENUS: *Nephelium L.* SPECIES: *Nephelium mutabile* Blume.

SELECTING A TREE. When selecting female and male trees from the nursery you should check the overall health of the trees. Look for active growth,

Unripe pulasan on a tree *(Tu7uh/Wikimedia Commons)*

such as new or young leaves. Make sure their leaves are not yellow or discolored, and free from insects or signs of insect damage. Check the trees to make sure they are free from injury, such as broken branches, cuts in the trunk, etc. And if possible, gently lift the trees out of their pots to check the roots. Overgrown trees that have seriously girdled roots will likely develop a poor root structure so they should be avoided.

How to Grow Pulasan Trees. Pulasan trees are tropical trees preferring an average temperature of 82 degrees Fahrenheit, high humidity, and an annual rainfall of 78 to 118 inches. The trees grow in sandy, loam and clay soils providing there adequate drainage. It prefers soil with organic matter and a pH of 5.0 to 5.8, but can grow in soil with a pH up to 6.5.

Select a sunny location that can accommodate the size of mature trees. In fields, the trees are generally planted nine to 12 yards apart. This usually isn't possible in many home backyards. But bear this in mind when selecting a site to plant your trees. Dig a hole at least twice the diameter of the pot the tree came in, about as deep as the container. Place the tree in the hole, loosening or cutting away any girdled roots. The tree should be planted so that the top of the root ball is about one to two inches higher than the soil level. Refill the hole with the original soil. Water the newly planted tree thoroughly. The trees like moist soil. The tree will benefit from the application of a compound fertilizer. Tree pests include ants, aphids and mealy bugs.

How to Grow from Seed. Pulasan seeds are only viable for a very short period, so they should be planted shortly after being removed from the fruit. Once you remove the seed, wash it under water and remove any pulp away from the seed. Immediately plant the seed in well draining soil rich in organic matter. The seed germinates best in temperatures in the low 80s. Keep the soil moist. Note that seeds from ripe pulasans germinate quickly.

▶ USDA Hardiness Zones: 11–12. Chilling Hours: 0. Water Requirements: 78–118 inches rainfall annually.

How to Consume. Pulasans are usually eaten fresh: Simply grab both ends of a fresh pulasan and twist it open. Or you can use a knife to cut it in half. You can discard, plant, or eat the seed in the center. Pulasans are also frozen for consumption later. The raw seeds have a taste similar to almonds, and can also be roasted. The seeds are also boiled to make a warm cocoa like beverage.

Additional Information. The fruit's English name comes from the Malaysian word "pulas," which means to twist. The fruit is often twisted with both hands to open it to expose the pulp for consumption. The pulasan's leaves and bark contain hydrocyanic acid.

References

Belen, Edwin. "Pulasan." *Archives of the Rare Fruit Council of Australia.* May 1981. http://rfcarchives.org.au/Next/Fruits/Pulasan/Pulasan5-81.htm. Accessed 09/06/18.
Fern, Ken. "Nephelium ramboutan-ake." Tropical Plants Database. http://tropical.theferns.info/viewtropical.php?id=Nephelium+ramboutan-ake. Accessed 09/06/18.
Galacgac, Evangeline S., and Criselda M. Balisacan. "Traditional Weather Forecasting Methods in Ilocos Norte." *Philippine Journal of Crop Science*, vol. 26, no. 1, 2001, p. 5–14. www.cabi.org/GARA/FullTextPDF/2009/20093019295.pdf. Accessed 12/30/18.
"Hairless Rambutan." Tropical Fruit Farm. www.tropicalfruits.com.my/pdf/Hairless-rambutan-Pulasan-k.pdf. Accessed 09/06/18.
Love, Ken, et al. "Tropical Fruit Tree Propagation Guide." F_N-49. College of Tropical Agriculture and Human Resources, University of Hawaii at Manoa, July 2017. www.ctahr.hawaii.edu/oc/freepubs/pdf/F_N-49.pdf. Accessed 09/06/18.
Mojica, Miko Jazmine J. "Weather lores have scientific basis." *Bureau of Agricultural Research and Development Digest* (Republic of the Philippines), vol. 7, no. 1, January–March 2005.

www.bar.gov.ph/index.php/digest-home/digest-archives/77-2005-1st-quarter/4453-jan mar05-weather-lores. Accessed 12/27/18.

Morton, Julia F. *Fruits of Warm Climates.* Eugene, OR: Wipf and Stock Publishers, 2003.

"*Nephelium mutabile* Blume." USDA, NRCS. 2018. The PLANTS Database (http://plants. usda.gov). National Plant Data Team, Greensboro, NC 27401-4901 USA. https://plants.sc. egov.usda.gov/core/profile?symbol=NEMU5. Accessed 09/08/18.

"*Nephelium ramboutan-ake* (Labill.) Leenh." National Parks Flora & Fauna Web, National Parks Board, Singapore. https://florafaunaweb.nparks.gov.sg/Special-Pages/plant-detail. aspx?id=3043. Accessed 09/06/18.

"Nephelium rambutan-ake-(Labill.) Leenh." Plants for a Future. https://pfaf.org/User/ Plant.aspx?LatinName=Nephelium+ramboutan-ake. Accessed 09/06/18.

Orwa, C., et al. "Nephelium ramboutan-ake." 2009 Agroforestree Database: A Tree Reference and Selection Guide Version 4.0. www.worldagroforestry.org/treedb/AFTPDFS/Nephelium_ rambutan-ake.PDF. Accessed 09/06/18.

Popenoe, Wilson. *Manual of Tropical and Subtropical Fruits: Excluding the Banana, Coconut, Pineapple, Citrus Fruits, Olive and Fig.* NY: Macmillan Co., 1920.

Pulasan—Sweeter Than Honey. Kanjirappally, India: Homegrown Biotech. n.d. http:// homegrown.in/pdf/new/PULASAN.pdf. Accessed 09/06/18.

Subhadrabandhu, Suranant. *Under-Utilized Tropical Fruits of Thailand.* RAP Publication: 2001/26. Bangkok: Food and Agriculture Organization of the United Nations, December 2001.

Purple Mangosteen (*Garcinia mangostana* L.)

Commonly known as just a mangosteen, and the Queen of Tropical Fruit, when ripe it has a smooth, hard, deep red-purple outer rind with a prominent calyx. The size of an orange, or a small apple, it contains a yellowish latex pericarp with a purple juice that stains upon contact. The edible fragrant inner pulp is soft, white, segmented and slightly acidic. Its flavor is frequently described as a cross between a peach and a pineapple or citrus. But its taste has also been described as a cross between a lychee, peach, strawberry and pineapple. Regardless of how its flavor is described, it is highly praised for its delicious unique tropical fruit flavor.

Fresh mangosteens shipped to mainland United States from Hawaii or Asia are required to be irradiated to prevent the potential of hitch-hiking fruit flies or other pests from entering the country. The irradiation supposedly has no noticeable impact on the fruit's taste. However, some mangosteen connoisseurs disagree and often seek out mangosteens from Puerto Rico which are not irradiated. Despite its name, mangosteens are not related to mangos.

Purple Mangosteen Lore. An unsubstantiated story has been passed down through the ages regarding Queen Victoria and the mangosteen: hearing of its fabulous delicious taste, Queen Victoria offered a small fortune and even knighthood to anyone who could deliver a fresh ripe mangosteen to her for her consumption.

In Asian folk medicine, mangosteens have been used to treat a variety of conditions and to boost the immune system. In some countries the rind is applied externally to treat acne and eczema. The mangosteen rind is also boiled and drunk as a tea to treat diarrhea and gonorrhea. The rind has also been dried and powdered and ingested to treat dysentery. Some sellers also claim it helps fight cancer, and improves brain function. However, there is presently an absence of scientific studies on the benefits of mangosteens on humans to substantiate its safety and effectiveness in treating any of the above conditions.

TREE HISTORY. Originating in the Sunda Islands (a group of islands in Indonesia and Malaysia), it then made its way to the Phillipines, Burma and India. It was introduced in England in 1789, and in Australia in 1854. And in 1855 the first mangosteen fruit were harvested from trees grown in green-houses in England. It was dubbed the "Queen of Tropical Fruits" in 1903 by David Fairchild, a famous horticulturalist.

Mangosteen seeds were received by the U.S. Department of Agriculture in 1906. Commercial attempts were made to grow mangosteens in Hawaii, Florida and California, but have been basically unsuccessful in Florida and California, since the trees do not acclimatize well. The long period before the trees begin producing fruit was also a factor.

Mangosteens were being studied at the Agricultural Research Service's

Purple mangosteen *(Jeremy Weate from Abuja, Nigeria/Wikimedia Commons)*

Cut purple mangosteen *(Basile Morin/Wikimedia Commons)*

Tropical Agriculture Research Station in Puerto Rico in 2008 where scientists are trying to accelerate its growth rate. Mangosteens are grown commercially to a very limited extent in Puerto Rico. Unfortunately 2017's hurricane Maria destroyed a lot of the orchards on the island.

TREE CHARACTERISTICS. Mangosteen trees are evergreen tropical trees. Under ideal conditions it grows up to 30 feet tall. Its canopy is somewhat pyramid shaped. The trees have dark green oblong leaves measuring up to 10 inches long. The leaves are glossy on top and dull on their under side. The tree's bark is dark brown, almost black. The tree bears large pink blossoms measuring up to two inches wide. The blossoms appear in either pairs or clusters depending upon whether they are male, or both male and female. The trees are slow growing trees. They take eight to ten or more years before they begin bearing fruit. Mature trees can grow up to 1000 fruit annually.

The trees are primarily propagated by seed since other propagation methods are generally not successful. However, it was reported that in Florida that they had some success with grafting *Garcinia mangostana* to seedlings of *Garcinia tinctoria* and allowing the plant to grow on both root systems. And in Puerto Rico mangosteens (*Garcinia* mangostana) have been successfully grafted onto root stocks of *Garcinia venulosa*, *Garcinia xanthochymus* and *Garcinia hombroniana*.

Mangosteen seeds are apomictic. Apomixis is the asexual production of seeds, meaning viable seeds are produced without pollination or fertilization. Therefore plants grown from apomictic seeds are clones of the original plant.

▶ FAMILY: Clusiaceae/Guttiferae. GENUS: *Garcinia* L. SPECIES: *Garcinia mangostana* L.

SELECTING A TREE. If you a purchasing a tree, understand they have delicate taproots. Select a tree at least two years old and two feet tall to insure it has a developed taproot, or else the tree will be too delicate to transplant into the ground. Check the tree's overall health. Look for active growth, such as new or young leaves. Make sure its leaves are not yellow or discolored, and free from insects or signs of insect damage. Check the entire tree to make sure it is free from injury, such as broken branches, cuts in the trunk, etc.

HOW TO GROW A PURPLE MANGOSTEEN TREE. The tree is a tropical tree, meaning it thrives in warm humid areas. It doesn't tolerate temperatures below 40 degrees Fahrenheit or over 100 degrees Fahrenheit. They require a minimum of 50 inches of rainfall annually. They will not grow in drought prone areas. Trees should be planted in well draining soil rich with organic matter. The tree is not adapted to limestone or sandy alluvial soils. Trees need great care when transplanting into the soil because they have extremely delicate tap roots. Newly planted trees should also be protected from strong winds.

Select a sunny location that can accommodate the size of a mature tree. Dig a hole at least twice the diameter of the pot the tree came in, about as deep as the container. Place the tree in the hole, loosening or cutting away any girdled roots. The tree should be planted so that the top of the root ball is about one to two inches higher than the soil level. Refill the hole with the original soil. Water the newly planted tree thoroughly. Young trees should be fertilized every quarter during their first three years. After that, they need only be fertilized once or twice a year.

If you live on the mainland of the United States, and you have a greenhouse, you may want to experiment with growing your mangosteen tree in a greenhouse where the temperature and humidity can be controlled.

HOW TO GROW FROM SEED. To grow from seed, remove the seeds from a fresh mangosteen and clean it with water. If you don't plan on immediately planting the seed, it needs to be kept moist. This can be done by storing it in a damp paper towel or peat moss, placed in a plastic bag in the refrigerator for a few days. Be aware that a study done at Kasetsart University (Bangkok, Thailand) showed that seeds stored over seven days before planting had a lower germination rate, and for those seeds stored for 10 days the germination rate dropped to 50 percent. So for best success, seeds should be planted within a few days of removing them from the fruit. Soak the seeds in water for 24 hours before planting them in well draining soil enriched with organic matter. Water the soil. And if you planted the seeds in a pot as opposed to in the ground, be sure to place the pot in a sunny location. The seeds will germinate within three weeks.

Although most fresh mangosteens you find on the mainland have been irradiated, the level of irradiation they've been subjected to will result in significantly lower germination rates than those seeds not irradiated. The level of radiation fresh Mangosteens undergo is determined by whether they are being irradiated to kill fruit flies or thirps, depending upon where the fruit is being imported from and the specific pest risk. It should also be noted that some experiments have been done where mangosteen seeds have been exposed to varying levels of gamma rays to increase morphological variability. Research studies show that the higher the gray (gy) dose of radiation mangosteen seeds are subjected to, the greater the decrease in the tree's overall height, number of leaves, etc.

The major pest in Hawaii is a caterpillar, *Stictoptera cuculioides* Guenee (Lepidoptera: Noctuidae), that feasts on new leaves and shoot tips. The caterpillar can be controlled with insecticides containing a neem extract.

▶ USDA Hardiness Zone: 11. CHILLING HOURS: 0. WATER REQUIREMENTS: minimum 50 inches rainfall annually.

How to Consume. Fresh mangosteens do not ripen after harvest. To select a fully ripe mangosteen, look for one which has a deep red/purplish skin. To eat fresh, break or cut the shell in half (be careful not to let the purple juice stain your clothes). Pull the two halves apart to expose the white flesh. The flesh will be in segments like oranges. Carefully pull out the segments and enjoy. (Tip: The number of segments in each mangosteen corresponds to the number of lobes at the apex of the rind.) According to the University of Hawaii's College of Tropical Agriculture and Human Resources, fresh mangosteens can be stored at 54 to 57 degrees Fahrenheit for up to 20 days. Commercially, mangosteens are canned and mangosteen juice is also sold.

Additional Information. Mangosteens contain xanthones, which are powerful antioxidants. Some early studies of xanthones on animals show they hold promise for reducing inflammation, and the proliferation of cancer cells, as well as hindering the oxidation of bad LDL cholesterol.

You may have heard of African mangosteens (*Garcinia livingstonei*). They are not the same as Purple Mangosteens (*Garcinia mangostana*). They share the same plant genus, but are different species. The African mangosteen tree is widespread in warmer regions of Africa. African mangosteens measure up to one and a half inches in diameter and possess an orange rind and flesh.

REFERENCES

Apple, R.W., Jr. "Forbidden Fruit: Something About a Mangosteen." *New York Times*, Sept. 24, 2003, p. F1–F5.

Bauer, Brent. "Can Drinking Mangosteen Juice Reduce Arthritis Inflammation And Pain?" Mayo Clinic. https://www.mayoclinic.org/diseases-conditions/rheumatoid-arthritis/expert-answers/mangosteen/faq-20058517. Accessed 05/01/18.

Bin Osman, Mohamad, and Abd Rahman Milan. *Mangosteen: Garcinia Mangostana L.* Southampton Center for Underutilised Crops, University of Southampton, Southampton, U.K. 2006.

Brunner, Bryan R., and J. Pablo Morales-Payan. "Soils, Plant Growth and Crop Production—Mangosteen and Rambuttan." In: *Encyclopedia of Life Support Systems*. UNESCO. www.eolss.net. Accessed 05/02/18.

Campbell, Carl W. "Growing Mangosteen in South Florida." *Proceedings of the Florida State Historical Society*, 1966, v. 79, p. 399–401. https://fshs.org/proceedings-o/1966-vol-79/399–401%20(CAMPBELL).pdf. Accessed 05/01/18.

Duff, Diana. "Plant of the Month: Mangosteen." *West Hawaii Today*. June 8, 2017.

"An Exotic Fruits Sampler." *Agricultural Research*, vol. 56, no. 1, January 2008, p. 15.

"Fresh Talent." *Good Health* (Australia edition), April 2013, p. 90–92.

"*Garcinia livingstonei.*" South African National Biodiversity Institute. http://pza.sanbi.org/garcinia-livingstonei Accessed 05/09/18.

"*Garcinia mangostana* (Clusiaceae)." Montoso Gardens. http://www.montosogardens.com/garcinia_mangostana.htm. Accessed 05/01/18.

"*Garcinia mangostana* L." USDA, NRCS. 2018. The PLANTS Database. National Plant Data Team, Greensboro, NC 27401-4901 USA. https://plants.sc.egov.usda.gov/core/profile?symbol=GAMA10. Accessed 12/20/18. 05/11/18.

Homhuan, Maneerat. "Study on Viability and Germination of Mangosteen Seeds (*Garcinia Mangostana* Linn.)" Thai National Agris Centre. http://agris.fao.org/agris-search/search.do?recordID=TH2016000279. Accessed 05/02/18.

"How to Grow Mangosteen." https://plantinstructions.com/tropical-fruit/how-to-plant-mangosteen/. Accessed 05/01/18.

"Is Mangosteen a Miracle Fruit? Science Doesn't Seem to Support Claims." *Environmental Nutrition*, Vol. 32, no. 12, December 2009, p. 3.

Karp, David. "Forbidden? Not the Mangosteen." *New York Times*, Food Section, Aug. 9, 2006. www.nytimes.com/2006/08/09/dining/09mang.html. Accessed 05/11/18.

Karp, David. "The Most Delicious Fruit on Earth? Terrific Mangosteens This Weekend." *Los Angeles Times*, Aug. 30, 2013. www.latimes.com/food/dailydish/la-dd-mangosteen-most-delicious-fruit-on-earth20130829-story.html. Accessed 05/04/18.

Kuswadi, Achmad Nasroh, Murni Indarwatmi, Indha Arastuti Nasution, and Hadian Iman Sasmita. "Minimum Gamma Irradiation Dose for Phytosanitary treatment of *Exallomochlus hispidus* (Heemiptera: Pseudococcidae)." *Florida Entomologist*, vol. 99, special issue no. 2, 2016, p. 69–75.

Marley, Karin. "Exotic Fruit." *Maclean's*, vol. 117, no. 29, July 19, 2004, p. 51.

Morton, Julia F. *Fruits of Warm Climates*. Eugene, OR: Wipf and Stock Publishers, 2003.

Nagao, Mike A., Heather M. C. Leite, Arnold H. Hara, and Ruth Y. Niino-DuPonte. *Mangosteen Caterpillar*. Insect Pests IP-14. University of Hawaii at Manoa, College of Tropical Agriculture and Human Resources, Cooperative Extension. 2004.

Nagy, Steven, and Philip E. Shaw. *Tropical and Subtropical Fruits: Composition, Properties and Uses*. Westport, CT: AVI Publishing. 1980.

Namkoong, Joan. "The Queen of Fruit." *Honolulu Magazine*, Sept. 1, 2006. www.honolulumagazine.com/Honolulu-Magazine/September-2006/The-Queen-of-Fruit/. Accessed 05/06/18.

"Nature's Powerful Juices." *Total Health*. Vol. 28, no. 2, June/July 2006, p. 42–46.

Paull, Robert E., and Saichol Ketsa. "Mangosteen: Postharvest Quality-Maintenance Guidelines." Fruit, Nut, and Beverage Crops, F_N-31. College of Tropical Agriculture and Human Resources, University of Hawaii at Manoa, Cooperative Extension. September 2014. www.ctahr.hawaii.edu/oc/freepubs/pdf/F_N-31.pdf. Accessed 05/30/18.

Paull, Robert E., and Saichol Ketsa. *Mangosteen*. www.researchgate.net/publication/237374174. Accessed 05/04/18.

Piper, Jacqueline. *Fruits of South-East Asia: Facts and Folklore*. New York: Oxford University Press, 1989.

Production Guideline: African Mangosteen: Garcinia Livingstonei. Agriculture, Forestry and Fisheries Department, Republic of South Africa. 2016.

Robb-Nicholson, Celeste, M.D. "Ask the Doctor: Does Mangosteen Have Any Health Benefits?" *Harvard Woman's Health Watch*. May 2012. www.health.harvard.edu/healthy-eating/does-mangosteen-have-any-health-benefits. Accessed 05/01/18.

Schardt, David. "Squeezing Cold Cash Out of Three 'Hot' Juices." *Nutrition Action Newsletter*, November 2006, p. 9–11.

Sobir, and Roedhy Powerwanto. "Mangosteen Genetics and Improvement." *International Journal of Plant Breeding*, November 2007, p. 105–111.

Stone, Daniel. "Meet the Mangosteen." National Geographic, May 26, 2016. www.nationalgeographic.com/people-and-culture/food/the-plate/2016/05/meet-the-mangosteen/. Accessed 05/06/18.

Stuart, Armando Gonzalez. "Mangosteen." University of Texas, El Paso / Austin Cooperative Pharmacy Program and Paso del Norte Health Foundation. www.herbalsafety.utep.edu/herbal-fact-sheets/mangosteen/. Accessed 05/01/18.

Swain, Roger B. "In Search of the Mangosteen." *Horticulture*, vol. 69, no. 10, December 1991, p. 54.

"Tropical Delight." *Burke's Backyard*, May 2009, p. 70–71.

Turner, Lisa. "Mangosteen: The Miracle Fruit?" *Better Nutrition*, vol. 68, no. 11, November 2006, p. 18–19.

Widiastuti, A., Sobir, and M.R. Suhartanto. "Diversity Analysis of Mangosteen (*Garcinia mangostana L.*) Irradiated by Gammaray Based on Morphological and Anatomical Characteristics." *Nusantara Bioscience*, vol. 2, no. 1, March 2010, p. 23–33.

www.mangosteen.com website. Accessed 08/26/17.

Rambutan (*Nephelium lappaceum* L.)

Also known as the hairy fruit, the fruit's name is derived from the Malay word "rambut" which translates into "hairy." If you're a science fiction film fan, rambutans look like miniature menacing alien creatures from a '50s B film. Rambutans are small round or oval fruit depending upon the cultivar, with soft green long fleshy spikes covering its hard skin that turns bright red or yellow when mature. On average, the fruit measure about one and three-quarter inches long to one and a half inches wide.

The interior flesh is juicy, translucent white with a single large seed in the center. The flesh clings to the seed, although a number of "freestone" cultivars exist. It has a sweet mildly acidic flavor and tastes something like a grape.

RAMBUTAN LORE. In Malaysia there's a folktale about a young girl choosing a man to be her husband by his fearful mask capped by a curly wig. She

Rambutan *(USDA Agricultural Research Service/Wikimedia Commons)*

knows behind the mask she will find a handsome, kind man. Just like a rambutan, the girl knows that beneath the unsightly exterior lies a promising delicious interior.

Although there are new cultivars where the rambutan flesh does not stick to the seed, it's said that long ago ladies of the Thai court were trained to be skilled in removing all traces of flesh from the seed using a sharp curved knife. They would then replace the flesh in the skin so that it appeared the flesh had been untouched. Poultices made from rambutan leaves are supposed to cure headaches. In Malaysia a concoction made from the tree roots is said to cure a fever, and the bark is supposedly a cure for tongue diseases.

TREE HISTORY. Rambutan trees originated in Malaysia and are grown throughout Southeast Asia. Arab traders brought the rambutan to East Africa sometime between the 13th and 15th centuries. Rambutans were subsequently introduced into India and South and Central America. In 1906 rambutan seeds were imported into the United States from Java. The rambutan made its way from Indonesia to the Philippines in 1912. They were introduced into

Australia in the 1930s. Rambutan trees are grown in large numbers in Thailand, Indonesia, Malaysia, India, and Australia. They are also grown in other countries with a tropical climate but to a lesser extent.

TREE CHARACTERISTICS. A tropical evergreen tree, a rambutan tree can grow 50 to 80 feet tall. Its trunk is erect, measuring up to two feet wide, with slightly creased grayish or reddish bark. The tree has a dense spreading crown. Its leaves are alternate, elliptic, measuring from about two inches up to a foot long. The tree bears small whitish, yellowish or greenish petal-free blossom clusters. Blossoming is stimulated by water stress. The resulting fruit grows in loose clusters. Rambutan trees are either male, female or hermaphroditic (producing both male and female blossoms). Male trees will not produce fruit. Female trees bear blossoms which require pollination to bear a decent fruit crop.

Several dozen cultivars of rambutan trees exist. Sample popular cultivars include the 'Maharlika' (pearly white flesh, medium juicy, subacid to sweet), 'Seematjan' (pearly white flesh, medium juicy to slightly dry, very sweet), 'Seenjojan' (translucent flesh, very juicy), 'Rongrein' (pearly white flesh, good flavor, easily separates from seed), and 'Sri Chompoo' (good flavor, easily separates from seed). The trees can live up to 60 years. However, their economic life is only 15 to 20 years, possibly up to 30. The trees generally fruit twice a year, with five to seven year old trees bearing 1000–1500 fruit per tree annually, with older trees producing 5000–6000 fruit annually.

Trees are propagated by seed, and vegetatively (cleft grafting, patch budding, marcottage). If the trees grown from seed are female or hermaphroditic, they usually begin fruiting in five to six years. Budded trees tend to begin fruiting in two to three years. Grafted trees also tend to bear larger fruit than trees grown from seed.

▶ FAMILY: Sapindaceae. GENUS: *Nephelium* L. SPECIES: *Nephelium lappaceum* L.

SELECTING A TREE. Although the number of cultivars available may be limited to you, since growing a rambutan tree is a time and financial investment, make sure you are purchasing a cultivar you desire. Remember there are both sweet cultivars for eating fresh, and some sour ones for stewing. At a nursery, check the tree's overall health. Look for active growth, such as new or young leaves. Make sure its leaves are not yellow or discolored, and free from insects or signs of insect damage. Check the entire tree to make sure it is free from injury, such as broken branches, cuts in the trunk, etc. And if possible, gently lift the tree out of the pot to check the roots. Overgrown trees that have seriously girdled roots will have a poor root structure so they should be avoided.

How to Grow a Rambutan Tree. Rambutan trees grow in a wide variety of soils providing there is good drainage. However, they grow best in deep clay loam or sandy loam rich in organic matter, with a soil pH between 4.5 and 6.5. They require regular irrigation.

Select a sunny location or one with partial shade shielded from the wind, since wind can damage fruit blossoms and ultimately fruit production. Make sure you select a planting site that can accommodate the size of a mature tree. Dig a hole at least twice the diameter of the pot the tree came in, about as deep as the container. Place the tree in the hole, loosening or cutting away any girdled roots. The tree should be planted so that the root ball is about one to two inches higher than the soil level. Refill the hole with the original soil. Water the newly planted tree thoroughly. Remember these are tropical trees that prefer humidity and do not tolerate drought. Trees should be fertilized to promote growth. An application of an NPK fertilizer of 6:6:6 or 8:8:8 is generally recommended three time a year up to year four, and then only an application twice a year after that.

Rambutans are a host to 118 different species of insects. If you have an insect problem with your rambutan tree, your local Cooperative Extension office can help you identify the insect and provide information on controlling or eradicating the pest. Since the fruit grows in clusters, sometimes numbering in the dozens, branches can sag under the weight of ripening fruit. Harvest the rambutan by cutting off the entire cluster.

How to Grow from Seed. Remove the rambutan seed from the fruit and wash it thoroughly. The seed should be planted immediately since the germination rate for seeds removed from fruit after a week drops to 50 percent. Plant the seed horizontally, with the flattened side down in a pot with drainage holes filled with soil amended with compost and sand. The seed should germinate in 10 to 21 days. Young seedlings need to be protected from the wind. They should not be planted outdoors until they are about two years old and hardy enough.

Be aware that if you trying to grow a tree from seed from rambutans imported from Thailand, the fruit was required to be irradiated prior to entering the country. Irradiation significantly lowers the germination rates of the seeds. If you are trying to grow a tree from seed, try to secure seeds from fresh rambutans grown in the United States, although it may be difficult since rambutans are produced commercially on only a very small scale in Hawaii.

► USDA Hardiness Zones: 10–11. Chilling Hours: 0. Water Requirements: Considerable.

How to Consume. To eat fresh, select ripe fruit with a bright red, or yellow skin depending upon the cultivar, with no signs of rot or decay at the stem end. Use a knife to gently cut around the fruit. Pull the two halves apart. You can now squeeze the flesh out of the skin, or scoop it out using a spoon. Be sure to discard the seed. Fresh rambutans bruise easily, so handle them with care. Rambutans can be added to fresh and green salads. They also make a nice garnish for Asian meat dishes. Fresh rambutans are also muddled in alcoholic cocktails. Fresh rambutans can be stored at 46 to 59 degrees Fahrenheit with 90 to 95 percent humidity for 14 to 16 days. Fruit refrigerated at 41 degrees Fahrenheit in an airtight container can be stored for up to three weeks.

ADDITIONAL INFORMATION. The rambutan seed, seed covering and fruit rind contain the toxins saponin and tannin. An oil produced by heating the seeds has sometimes been used in making soap and candles.

REFERENCES

Diczbalis, Yan. *Rambutan Improving Yield and Quality.* Kingston, ACT: Rural Industries Research and Development Corporation. 2002. https://rirdc.infoservices.com.au/down loads/02-136.pdf. Accessed 08/30/18.

Galindo, R.A., and M.P. Loquias. *Rambutan Production Guide.* Bureau of Plant Industry, Department of Agriculture, Philippines. n.d. http://bpi.da.gov.ph/bpi/images/Production_ guide/pdf/RAMBUTAN.pdf. Accessed 09/03/18.

Goebel, Roger. "Rambutan Cultivation." *Archives of the Rare Fruit Council of Australia*, September 1988. http://rfcarchives.org.au/Next/Fruits/Rambutan/RambutanPropagation9-88.htm. Accessed 09/03/18.

Grant, Amy. "Rambutan Growing Tips: Learn About Rambutan Tree Care." Gardening Know How. April 4, 2018. www.gardeningknowhow.com/edible/fruits/rambutan/rambutan-growing-tips.htm. Accessed 08/30/18.

Ketsa, Saichol, and Robert E. Paull. *Rambutan: Postharvest Quality-Maintenance Guidelines.* F_N-33. Cooperative Extension, College of Tropical Agriculture and Human Resources, University of Hawaii at Manoa. May 2014. www.ctahr.hawaii.edu/oc/freepubs/pdf/F_N-33.pdf. Accessed 08/29/18.

Morton, Julia F. *Fruits of Warm Climates.* Eugene, OR: Wipf and Stock Publishers, 2003.

Muhamed, Sameer, and Sajan Kurian. "Phenophases of Rambutan (*Nephelium lappaceum* L.) based on extended BBCH-scale for Kerala, India." *Current Plant Biology*, vol. 13, 2018, p. 37–44.

"Nephelium lappaceum." EcoCrop Database. United Nations, Food and Agriculture Organization. http://ecocrop.fao.org/ecocrop/srv/en/cropView?id=1528. Accessed 09/03/18.

"*Nephelium lappaceum* L." USDA, NRCS. 2018. The PLANTS Database. https://plants.usda. gov/core/profile?symbol=NELA7. Accessed 09/01/18.

Ortho All About Citrus and Subtropical Fruits. Des Moines, IA: Meredith Books, 2008.

Orwa, C., et al. "Nephelium lappaceum." 2009 Agroforestree Database: A Tree Reference and Selection Guide Version 4.0. www.worldagroforestry.org/treedb/AFTPDFS/Nephelium_ lappaceum.PDF. Accessed 08/30/18.

Piper, Jacqueline M. *Fruits of South-East Asia: Facts and Fiction.* Oxford, NY: Oxford University Press, 1989.

Popenoe, Wilson. *Manual of Tropical and Subtropical Fruits: Excluding the Banana, Coconut, Pineapple, Citrus Fruits, Olive and Fig.* NY: Macmillan Co., 1920.

"Rambutan." Department of Health, Government of Western Australia. http://healthywa.wa. gov.au/Recipes/N_R/Rambutan. Accessed 08/30/18.

"Rambutan." Rambutan.com. http://www.rambutan.com/. Accessed 08/30/18.

"Rambutan." www.tropicalfruits.com.my/pdf/Rambutan-k.pdf. Accessed 05/27/18.

Schisandra Berry (*Schisandra chinensis*)

Also commonly known as Wu Wei Zi (Five Flavored Berry), the Five Flavor Fruit, and the Chinese Schizandra, the Schisandra berry turns deep red when ripe, and looks similar to a red currant. Measuring about a centimeter in diameter, the berries appear on the plant in grape-like clusters. The Schisandra berry is often referred to as the five flavored berry because of its distinct five flavors. Its taste is described as sweet, sour, bitter, salty and pungent. The berry's skin is said to provide the sweetness, the pulp the sourness and the juice saltiness, and the seeds the bitterness and pungency. The fruit is slightly acidic.

SCHISANDRA BERRY LORE. The Schisandra berry has been used in China to treat a variety of conditions, including physical and mental stress, fatigue, coughs, and liver conditions. It's also been used in herbal medicine to prevent aging, and to promote stamina and endurance. It's rumored to normalize blood sugar. The Schisandra berry is also used in Russian medicine. In Chinese herbal medicine, the Schisandra berry is frequently used like the gingko. Numerous studies have been and are being conducted on the effectiveness of Schisandra berry for treatment of a variety of health issues.

PLANT HISTORY. The Schisandra berry plant originated in China and Mongolia, and possibly the Russian Far East. It's native to the forests there and has been valued by the locals for over 2000 years. It was introduced into the United States in the late 1850s from Russia. The plant is now naturalized in Russia, Korean and Japan. It is also grown in North America, and in several European countries primarily as a landscape plant.

PLANT CHARACTERISTICS. The plant, also known as a Chinese Magnolia Vine because of its small fragrant magnolia like blossoms, is a deciduous woody climber. Although its vines can grow up to 30 feet in length, most plants reportedly generally only grow to about half that size. The vines feature alternate, elliptic-shaped, fragrant, shiny dark green leaves measuring approximately one to two inches wide and two to two and three-quarter inches long. In the spring the plant bears small clusters of three to five fragrant cream colored blossoms. The blossoms are pollinated by insects. The plant later produces crimson colored berries.

The plants can be propagated by seed, by root and softwood cuttings, and by air layering. Schisandra berry plants usually start to fruit in three to four years. The plants generally live 20 or more years. The plants are dioecious, meaning plants have either male or female blossoms. So you will need

Schisandra berry *(Vladimir Kosolapov/Wikimedia Commons)*

to grow both a male and female plant for cross pollination if you want them to bear fruit. A self fruitful cultivar does exist.

The plant vines are usually supported by a trellis, fence or other support system. Because of the plant's attractive flowers and ensuing grape-like clusters of berries, the plant is sometimes grown as a landscape plant.

▶ FAMILY: Schisandraceae. GENUS: *Schisandra*. SPECIES: *Schisandra chinensis.*

SELECTING A PLANT. In general, you will need to purchase and plant both a female and male plant if you want berries. If you decide to grow multiple plants, one male plant is needed for every three female plants. However, a self-fertile cultivar called the 'Eastern Prince' was developed in Russia. Some reports though say the 'Eastern Prince' sometimes reverts back to a dioecious plant.

Schisandra berry tea *(abex/flickr.com)*

How to Grow a Schisandra Berry Plant. Schisandra berry plants are hardy down to minus 40 degrees Fahrenheit. They grow best in a location that gets full morning sun and indirect afternoon sun, i.e., partial shade. They thrive in rich, moist, well draining soils, preferring sandy loam to clay loam soils. Although they prefer a soil pH of 5.5 to 6.5, they will grow in soil with a pH from 5.0 to 8.0. You will need to plant both a male and female Schisandra berry plant if you are growing them for fruit production. Select a location with partial shade that can accommodate the mature size of your plants, and that can accommodate a support system for them. Planting them by a fence or wall will provide the support they need. Plants should be spaced at least three feet apart from one another.

Most Schisandra berry plants are sold as bare root plants. When you receive or bring home your bare root plants, you need to soak the plant roots in a bucket of water a minimum of two hours before planting them. While the roots are soaking, you can dig holes slightly deeper than the length of the plant roots, and twice the size of the root's diameter. When you are ready to plant them, be sure to spread out the plant roots if needed, and set the plant in the hole. Refill the hole with soil. The plant bud should be level with or slightly above the soil. Water the plant thoroughly. Water the plants regularly to help them get established. They are not drought tolerant and need consistent moisture. Just be careful not to over water them.

As a climbing vine, if your plants are not adjacent to a fence or wall, you will need to install a trellis or garden stakes to give them a supporting structure. As the vines grow, if you need to tie them to your support system, use flexible ties. (Note: Old pantyhose and other nylons make great flexible ties.)

HOW TO GROW FROM SEED. Growing the plant from seed is usually not the preferred method of propagation because of their highly variable levels of germination. To grow from seed, soak stored or packaged seeds in warm water for 12 hours. Scarify the seeds and plant them in moist soil indoors in the spring. If possible, allow them to grow in a greenhouse for the first two years before planting them outside.

However, according to the *Interactive Agricultural Ecological Atlas of Russia and Neighboring Countries*, Schisandra seeds need to be cold stratified for four months at two to five degrees Centigrade.

▶ USDA HARDINESS ZONES: 4–8 for fruit, up to 10 as a landscape plant. CHILLING HOURS: Required for fruit set. (No information is available on the required number of chilling hours.) WATER REQUIREMENTS: Consistent moisture needed.

HOW TO CONSUME. Although they are tart, ripe Schisandra berries can be eaten fresh. Fresh berries can also be added to salads and the pulp added to fruit smoothies. Fresh berries can also be made into juice, and can be used in preserves. The berries can also be dried. In Korea the berries are made into a liqueur. And in China the berries are used to make a wine.

ADDITIONAL INFORMATION. WARNING: The WebMd website warns that you should avoid taking Schisandra if you have epilepsy, gastroesophageal reflux disease, peptic ulcers or high blood pressure because it could worsen these conditions.

Like ginseng, Schisandra berry is considered an "adaptogen." In herbal medicine adaptogens are believed to help the body adapt to stress, increase resistance to diseases, and increase your energy and vigor. Young plant leaves are sometimes cooked and eaten as a vegetable. Dried plants leaves are sometimes used to make tea.

REFERENCES

Bailey, Roberta. "Grow Your Own Schisandra." *Maine Organic Farmer and Gardener*, Summer 2016. www.mofga.org/Publications/The-Maine-Organic-Farmer-Gardener/Summer-2016/Schisandra. Accessed 12/11/18.

Epp, Tracy. "More Power to You." *Vegetarian Times*, Issue no. 259, March 1999, p. 84.

Foster, Steven. "Schisandra: A Rising Star." *Mother Earth Living*, April/May 2000. www.mother
   earthliving.com/Health-and-Wellness/HERBS-for-HEALTH-SchisandraA-Rising-
   Star.aspx. Accessed 03/22/16.
Ko, Kam-Ming, Jun Yin, and Chuixin Qin. *Schisandra chinensis: An Herb of North Eastern
   China Origin*. Singapore: World Scientific Publishing, 2015.
"Magnolia Vine Care Guide." Edible Landscaping. http://ediblelandscaping.com/careguide/
   MagnoliaVine/. Accessed 12/09/18.
"Magnolia Vine (Five Flavor Fruit)." Uncommon Fruit, Observations from the Carandale
   Farm, University of Wisconsin. http://uncommonfruit.cias.wisc.edu/magnolia-vine-five-
   flavor-fruit-schisandra-chinensis/. Accessed 12/08/18.
"Schisandra." WebMd. www.webmd.com/vitamins/ai/ingredientmono-376/schisandra.
   Accessed 12/08/18.
"Schisandra chinensis (Chinese Magnolia Vine)." Backyard Gardener. www.backyardgardener.
   com/plantname/schisandra-chinensis-chinese-magnolia-vine/. Accessed 12/08/18.
"*Schisandra chinensis* (Turcz.) Baill." Biodiversity Education and Research Greenhouses, Uni-
   versity of Connecticut. http://florawww.eeb.uconn.edu/200700132.html
"Schisandra chinensis (Turcz.) Baill.—Chinese magnolia vine." Interactive Agricultural Ecological Atlas of Russia and Neighboring Countries. http://www.agroatlas.ru/en/content/
   related/Schisandra_chinensis/index.html. Accessed 12/10/18.
"Schisandra chinensis—(Turcz.) Baill." Plants for a Future. https://pfaf.org/user/plant.aspx?
   latinname=Schisandra+chinensis. Accessed 12/08/18.
"Schisandra Information—How to Grow Schisandra Magnolia Vines." Gardening Know How.
   https://www.gardeningknowhow.com/ornamental/vines/schisandra/schisandra-magnolia-
   vines.htm. Accessed 12/08/18.
"Schisandra: Ultimate Superberry." Medicine Hunter. www.medicinehunter.com/schisandra.
   Accessed 12/09/18.
Schloemann, Sonia. "Project Title: Cultural Requirements for Cultural Production of Schisandra chinensis, Wu Wei Zi or Chinese Magnolia Vine." University of Massachusetts Extension. https://ag.umass.edu/sites/ag.umass.edu/files/fact-sheets/pdf/schisandra_project_
   report.pdf. Accessed 12/08/18.
Schmerler, Sam. "Plainly Unique: *Schisandra chinensis*." *Arnoldia: A Continuation of the Bulletin of Popular Information of the Arnold Arboretum, Harvard University*, vol. 69, no. 3,
   2012. http://arnoldia.arboretum.harvard.edu/pdf/articles/2012-69-3-plainly-unique-
   schisandra-chinensis.pdf. Accessed 12/08/18.
Stremple, Barbara Ferguson, editor. *All About Growing Fruits, Berries and Nuts*. San Ramon,
   CA: Ortho Books. 1987.
Szopa, Agnieszka, Radoslaw Ekiert, and Halina Ekiert. "Current Knowledge of *Schisandra
   chinensis* (Turcz.) Baill. (Chinese Magnolia Vine) as a Medicinal Plant Species: A Review
   on the Bioactive Components, Pharmacological Properties, Analytical and Biotechno-
   logical Studies." *Phytochemistry Reviews*, vol. 16, no. 2, 2017, p. 195–218.
Tyler, Varro E. "Liver-Loving Herbs." *Prevention*, vol. 53, no. 12, December 2001, p. 107–109.
Wikman, Georg, and Alexander Panossian. "Pharmacology of Schisandra chinensis Bail: An
   Overview of Russian Research and Uses in Medicine." *Journal of Ethnopharmacology*, vol.
   118, no. 2, July 23, 2008, p. 183–212.
"Wu Wei Zi. Chinese Magnolia Vine (Schisandra chinensis)." Dave's Garden. https://daves
   garden.com/guides/pf/go/82282/. Accessed 12/09/18.

Wampee (*Clausena lansium* [Lour.] Skeels)

Also known as a wampi, the wampee is a round or oblong fruit up to a little
more than an inch long and up to an inch in diameter. Its thin but tough skin
is slightly hairy and brown, which turns a yellowish brown when ripe. The
skin is not consumed. Depending upon the cultivar, the flesh is either white

Wampee *(WingkLEE/Wikimedia Commons)*

or pale yellow with a grape like texture. Each fruit contains one to five inedible large green oblong seeds.

There are sweet, subacid and sour cultivars. Their taste varies depending upon the cultivar but there seems to be no common description of their taste. The subacid cultivars are sometimes described as citrusy, the sweet cultivars as somewhat peachy, and the sour cultivars something like an unripe loquat.

Samples of sour cultivars include the 'Cows Kidney' ('Niu Shen'), 'Wild Growing' ('Yeh Sheng'), 'Sour jujube' ('Suan Tsao'), 'Long-chicken-heart-sour-wampee' and 'Yellow-hairy-chicken-heart-sour-wampee.' Sample sweet cultivars include the 'Chicken heart' ('Chi Shin') and the 'White-hairy-chicken-heart-sweet-wampee,' 'Guy Sam' and 'Yeem Pay.' Sample sub-acid cultivars include the 'Melon section' ('Kua Pan') and the 'Small chicken heart' ('Hsiao Chi Hsien').

WAMPEE LORE. In Asian folk medicine, wampees have been used as a treatment for hepatitis, bronchitis, inflammation, cancer and Parkinson's disease, among other health conditions. Wampees are not medically approved as a treatment for these health problems in Western medicine. It should be noted though that a study on the pharmacological effects of wampee bark reported in the *Journal of Ethnopharmacology* concluded, "The hepatoprotective activity of C. lansium is partly due to its anti-oxidant and anti-inflammatory properties and confirms its folkloric use in the treatment of gastro-intestinal inflammation, bronchitis and hepatitis." And an article in the journal *Acta Pharmacologica Sinica* reported on a study on Bu-7, a flavonoid extracted from wampee. That article concluded, "Bu-7 protects PC12 cells against rotenone injury, which may be attributed to MAP kinase cascade (JNK and p38) signaling pathway. Thus, Bu-7 may be a potential bioactive compound for the treatment of Parkinson's disease." And lastly, an article appearing in the *Journal of Biomedicine and Biotechnology* on the antioxidant and anticancer activities of wampee peel concluded, "...seems that the wampee peel extract can be used as natural antioxidant and anticancer agent. Further investigation is being carried out to identify and characterize the inherent phenolic compounds responsible for the antioxidant and anticancer activities from the ethylacetate fraction of wampee peel." More studies on wampee are underway.

TREE HISTORY. Originating in China, they spread to other regions in Southeast Asia in the 1700s. It's believed that Chinese migrants brought the wampee to other Asian countries such as Malaysia and Singapore. Wampees have been growing in India since the 1800s. Wampees were introduced into Florida and Hawaii in the early 1900s. The trees have been growing in Puerto Rico and Jamaica since the early 1900s. The tree is also grown in Australia.

TREE CHARACTERISTICS. A distant relative of the citrus family, wampee trees are evergreens reaching only 15 to 20 feet in height. It features upward slanting flexible branches. Its bark is rough and gray-brown. The tree's evergreen leaves are smooth, dark green, elliptic ovate and two and three-quarter to four inches long. A self pollinating tree, it bears fragrant white or yellowish-green four or five petaled blossoms. Each blossom is about half an inch wide and is borne in flower clusters from four to 20 inches long. Wampee fruit are produced in large grape like clusters. They dangle on quarter- to half inch stalks. Mature trees can produce hundreds to thousands of wampees. Wampee trees can be propagated by various methods. It can be propagated by air layering, by softwood cuttings, and by grafting. It can also be propagated by seed. Trees grown by seed begin bearing fruit in five to eight years.

▶ FAMILY: Rutaceae. GENUS: *Clausena* Burm. f. SPECIES: *Clausena lansium* (Lour.) Skeels.

SELECTING A TREE. The number of cultivars available may be very limited in your region. So make sure the tree you ultimately select produces your desired choice of a sweet, or sour, or subacid fruit. Be aware that if you live in California, wampee trees are not available commercially.

Check the tree's overall health. Look for active growth, such as new or young leaves. Make sure its leaves are not yellow or discolored, and free from insects or signs of insect damage. Check the entire tree to make sure it is free from injury, such as broken branches, cuts in the trunk, etc. Check to make sure the tree trunk is not sunburned. And if possible, gently lift the tree out of the pot to check the roots. Overgrown trees that have seriously girdled roots will have a poor root structure so they should be avoided.

How to Grow a Wampee Tree. Considered a subtropical to tropical tree, the tree tolerates a wide range of soils, providing there is adequate drainage. The trees are susceptible to root rot if they remain wet for extended periods. The tree grows best in rich loamy soils. They prefer full sun, but will tolerate partial shade. Select a sunny location or one with partial shade that can accommodate the size of a mature tree. Dig a hole at least twice the diameter of the pot the tree came in, about as deep as the container. Place the tree in the hole, loosening or cutting away any girdled roots. The tree should be planted so that the top of the root ball is about one to two inches higher than the soil level. Refill the hole with the original soil. Water the newly planted tree thoroughly. The tree should be watered regularly.

If a hard freeze is predicted for a period of time, you should protect the tree. You should wrap the tree trunk and branches in insulation material like cardboard, fiberglass or frost protection covers. Make sure the soil is moist and not dry, since damp soil retains and radiates more heat than dry soil. Bare soil also radiates more heat than soil covered with mulch or other ground covers. Remove the insulation during the day so that the tree can absorb the warm sunny daylight temperatures. If your tree does sustain frost damage, don't prune away any dead branches until the spring. This allows time for the tree to recover in warmer weather. It also allows you to better identify the damaged branches to be removed. Regular tree pruning is not required. Since the fruit is borne on the tips of branches, the less pruning the better. The only reported pests of the tree are aphids.

▶ USDA Hardiness Zones: 10–11. Chilling Hours: 0. Water Requirements: Average.

How to Consume. Wampee can be eaten fresh. Wash it under running water and pat it dry with a paper towel. Although the skin is edible, it's generally not consumed because it is resinous. So peel away the skin and consume the flesh, spitting out the inedible seeds in the fruit. Fresh wampees can be added to fruit salads, and to gelatins. The juice can be used in salad dressings and marinades, and to flavor frozen treats. They can be served alongside meat, or as a garnish. In Asia, wampees are sometimes fermented and made into a wine, or other alcoholic beverages.

ADDITIONAL INFORMATION. Wampees are high in vitamin C.

REFERENCES

Adebajo, A.C., et al. "Pharmacological Properties of the Extract and Some Isolated Compounds of Clausena lansium Stem Bark: Anti-Trichomonal, Antidiabetic, Anti-Inflammatory, Hepatoprotective and Antioxidant Effects." *Journal of Ethnopharmacology*, vol. 122, no. 1, February 25, 2009. p. 10–19.
Chiri, Alfredo. "Wampee—Clausena lansium var. 'Kua Pan'—Rutaceae." *California Rare Fruit Growers Los Angeles Chapter Newsletter*, vol. 8, no. 5, September 2005, p. 2.
"Clausena lansium." Citrus Variety Collection, University of California, Riverside. www.citrus variety.ucr.edu/citrus/wampee.html. Accessed 08/24/18.
"Clausena lansium." *Flora of China*, vol. 11, p. 83–84. www.efloras.org/florataxon.aspx? flora_id=2&taxon_id=200012437. Accessed 08/26/18.
"*Clausena lansium* (Lour.) Skeels." USDA, NRCS. 2018. The PLANTS Database. National Plant Data Team, Greensboro, NC 27401-4901. https://plants.sc.egov.usda.gov/core/profile? symbol=CLLA9. Accessed 12/26/18.
Fern, Ken. "Clausena lansium." Tropical Plants Database. http://tropical.theferns.info/view tropical.php?id=Clausena+lansium. Accessed 08/26/18.
Grant, Bonnie. "Wampi Plant Care—Growing an Indian Swamp Plant In Gardens." Gardening Know How. www.gardeningknowhow.com/edible/fruits/wampi/wampi-plant-care.htm. Accessed 02/07/18.
Li, B.Y., et al. "Protective Effect of Bu-7, a Flavonoid Extracted from Clausena lansium, Against Rotenone Injury in PC12 Cells." *Acta Pharmacologica Sinica*, vol. 32, no. 11, November 2011, p. 1321–1326.
Morton, Julia F. *Fruits of Warm Climates*. Eugene, OR: Wipf and Stock Publishers, 2003.
Ortho All About Citrus and Subtropical Fruits. Des Moines, IA: Meredith Books, 2008.
Prasad, K. Nagendra, et al. "Antioxidant and Anticancer Activities of Wampee (*Clausena lansium* (Lour.) Skeels) Peel." *Journal of Biomedicine and Biotechnology*, vol. 2009. http:// scholar.google.com/scholar_url?url=http://downloads.hindawi.com/journals/biomed/2009/ 612805.pdf&hl=en&sa=X&scisig=AAGBfm3cWgvRYgmcEneGApdu0DtMd7uERA& nossl=1&oi=scholarr. Accessed 08/27/18.
Vargo, Don. *Information on Some Tropical and Subtropical Fruit Trees at the Land Grant Agricultural Experiment Station, Malaeimi, Amerian Samoa.* Land Grant Technical Report No. 11. Pago Pago, American Samoa: Land Grant Program, Amerian Samoa Community College, American Samoa Government. October 1989.
"Wampee." Daleys Fruit Nursery. www.daleysfruit.com.au/fruit%20pages/wampee.htm. Accessed 02/07/18.
"Wampee Fruit." www.fruitsinfo.com/processed-products-of-fruits.php. Accessed 02/07/18. www.rarefruitclub.org.au/Level2/Wampi.htm. Accessed 02/07/18.
"Wampi, Also Spelled Wampee." Rare Fruit Club, Australia. "Wampee (*Clausena lansium*)." FGCU Food Forest Plant Database. Florida Gulf Coast University, Fort Myers, FL. www.fgcu.edu/UndergraduateStudies/files/Wampee.pdf. Accessed 08/24/18.

Yuzu (*Citrus junos*, synonym *Citrus ichangensis* × *Citrus reticulata* var. austera)

The yuzu citrus fruit has an average width of 1.9 inches and height of 1.7 inches. But the fruit can reach up to three inches in diameter. The yuzu's rind is dimpled and rough with prominent oil glands, and turns from green to yellow or yellowish orange when the fruit is mature. The yuzu is known for its unique strong fragrance. Its fragrance is sometimes described as a spicy grapefruit/lime smell.

Yuzu have a small amount of pulp and lots of inedible seeds, slightly over two dozen per fruit. The fruit is sour and highly acidic. Its taste is unique and thus has been described in many ways. Some describe the taste as a sweet orangey grapefruit, others describe it as tasting like a mild lemon with a touch of grapefruit bitterness, while others describe its flavor as a cross between a grapefruit and a lime. Due to its acidic and sour taste and numerous seeds, it is not consumed like an orange.

YUZU LORE. In Japan there's a tradition dating back to the 18th century involving the yuzu that some still practice today. During Toji, the winter solstice, some people would engage in a yuzuyu, or yuzuburo, a yuzu bath. Whole or half yuzus would be floated in warm bath water. The yuzu bath is believed to bring good fortune and good health, to relax the mind, and to protect against seasonal maladies like colds and dry skin.

TREE HISTORY. The tree originated in the Yangtze River area of China. It then made its way to Korea. From there it was introduced into Japan over 1000 years ago. The yuzu was introduced in the United States in 1914 by USDA explorer Frank Meyer who sent seeds to the United States.

Yuzu *(Nikita from Russian Federation/Wikimedia Commons)*

TREE CHARACTERISTICS. The yuzu is believed to be a natural hybrid of an Ichang papeda and a sour mandarin. A small evergreen tree or shrub, it usually grows up to six feet or taller. In its native habitat it can grow up to 18 feet tall. It bears small dark green leaves shaped like a lance head, and the branches have thorns. The tree bears very fragrant small white blossoms. It's different from other citrus trees in that the yuzu tree is cold hardy and can withstand temperatures down to 21 degrees Fahrenheit.

A number of cultivars exist, but many are not available in the United States. For example in Japan, one of the major tree producers, there's a sweet cultivar available, and another cultivar grown for its fragrant stunning blossoms rather than its fruit. Yuzu trees are commonly grafted, but can also be grown from seed and from pips. But note that trees grown from seed and from pips may not bear true to type. And trees grown from seed or pips can take up to 10 years before beginning to bear fruit.

▶ FAMILY: Rutaceae. GENUS: *Citrus.* SPECIES: *Citrus ichangensis × Citrus reticulata* var. austere.

SELECTING A TREE. At the nursery, check the tree's overall health. Look for active growth, such as new or young leaves. Make sure its leaves are not yellow or discolored, and free from insects or signs of insect damage. Check the entire tree to make sure it is free from injury, such as broken branches, cuts in the trunk, etc. And if possible, gently lift the tree out of the pot to check the roots. Overgrown trees that have seriously girdled roots will likely develop a poor root structure so they should be avoided.

HOW TO GROW A YUZU TREE. The tree grows in most soils providing it has adequate drainage. The tree also needs ample sun (at least six hours per day.) As with other citrus trees, the best time to plant a yuzu tree is in the spring when all danger of frost has past. This allows the tree time to establish itself before the winter. Select a sunny location or one with partial shade that can accommodate the size of a mature tree. Dig a hole at least twice the diameter of the pot the tree came in, about as deep as the container. Place the tree in the hole, loosening or cutting away any girdled roots. The tree should be planted so that the root ball is about one to two inches higher than the soil level. Refill the hole with the original soil. Water the newly planted tree thoroughly. The tree should be regularly watered the first two years.

Although the yuzu tree is a cold-hardy citrus tree, if a hard freeze is predicted for a period of time, you should protect the tree by wrapping the trunk and branches in insulation material like cardboard, fiberglass or frost protection covers. Make sure the soil is moist and not dry, since damp soil retains and radiates more heat than dry soil. Bare soil also radiates more heat than soil covered with mulch or other ground covers. Remove the insulation during the day so that the tree can absorb the warm sunny daylight temperatures. If your tree does sustain frost damage, don't prune away any dead branches until the spring. This allows time for the tree to recover in warmer weather. It also allows you to better identify the damaged branches to be removed. Young citrus trees benefit from fertilization. If using a citrus fertilizer, follow the instructions on the label or check with your local Cooperative Extension office for recommendations on fertilization.

How to Grow from Seed. Remove the seeds from the yuzu fruit. Wash any pulp away from the seeds. Soak the seeds in clean water overnight. Place the seeds about half an inch deep in a sterile moist planting/potting medium. Cover the pot with plastic to create a humid atmosphere. Keep the soil moist until seedlings sprout. The seeds germinate best in temperatures above 60 degrees Fahrenheit. The seeds usually germinate in three to four weeks. Once the seeds sprout, transplant them into a larger pot.

▶ USDA HARDINESS ZONES: 8–10. CHILLING HOURS: 0. WATER REQUIREMENTS: Moderate.

How to Consume. Because of the numerous seed, the juice and zest of yuzus are primarily used. Both are ingredients used in making Japanese Ponzu sauce, a popular soy-based sauce. Yuzu is also added to some miso soup recipes. Slivers of the rind are used to flavor dishes such as fish, noodles and cooked vegetables. It's used to add flavor to a variety of foods ranging from salad dressings to sorbet. In Korea, a yuzu marmalade is very popular. The juice is also added to water to make a refreshing citrus beverage. If you have a hard time finding fresh yuzus in local specialty markets, you can purchase bottled yuzu juice, marmalade and other products from online marketplaces.

Additional Information. WARNING: Yuzu can have an anticoagulant effect which could interact with blood thinners. So if you are taking a blood thinner like Warfarin or Coumadin, consult your doctor before consuming yuzu. Yuzu juice is being touted as a super juice because it has three times as much vitamin C as lemon juice. Yuzu peel "solvent" is used in various perfumes due to its pleasant scent.

A study reported in the *Journal of Alternative and Complementary Medicine* concluded, "Yuzu's aromatic effects may alleviate negative emotional stress, which, at least in part, would contribute to the suppression of sympathetic nervous system activity." Another study reported in the same journal "indicated that short-term inhalation of yuzu fragrance could alleviate premenstrual emotional symptoms."

Yuzus may potentially benefit your brain. A study on rats reported in the *Journal of Nutrition* that treatment with yuzu extract prevented cognitive dysfunction by reducing the buildup of beta amyloid proteins in the brain. The wood from the Yuzu is used to make the taepyeongso, the traditional Korean oboe.

REFERENCES

"Citrus : Perfume Lexicon & Fragrance Notes." Bois de Jasmin. July 24, 2005. https://boisde jasmin.com/2011/06/perfume-lexicon-fragrance-notes-all-about-citrus.html Accessed 08/17/18.

Embiricos, George. "Yuzu Is About to Explode in Popularity in the United States. Here's Why." *Food Republic*, October 28, 2014. www.foodrepublic.com/2014/10/28/yuzu-is-about-to-explode-in-popularity-in-the-united-states-heres-why/. Accessed 08/16/18.

Fabrikcant, Florence. "Citrus Fruits Take on New Appeal, Bring Fresh Blood Oranges, Yuzu to Attention." *Nation's Restaurant News*, vol. 33, no. 7, February 15, 1999, p. 48.

Geisel, Pamela M., and Carolyn L. Unruh. *Frost Protection for Citrus and Other Subtropicals.* ANR Publication 8100. University of California, Division of Agriculture and Natural Resources, 2003.

Kita, Paul. "It's Squeezin' Season!" *Men's Health,* vol. 27, no. 10, December 2012, p. 62.

Mahon, Stephanie. "How to Grow Yuzu." *The Telegraph*, January 14, 2016. www.telegraph.co.uk/gardening/how-to-grow/how-to-grow-yuzu/. Accessed 08/18/18.

Marano, Hara Estroff. "Fruit of the Future." *Psychology Today*, vol. 46, no. 5, September/October 2013, p. 47.

Matsumoto, Tamaki, et al. "Does Japanese Citrus Fruit Yuzu (*Citrus junos* Sieb. ex Tana.k.a.) Fragrance Have Lavender-Like Therapeutic Effects That Alleviate Premenstrual Emotional Symptoms? A Single-Blind Randomized Crossover Study." *Journal of Alternative and Complementary Medicine*, vol. 23, no. 6, 2017, p. 461–470.

Matsumoto, Tamaki, et al. "Effects of Olfactory Stimulation from the Fragrance of the Japanese Citrus Fruit Yuzu (*Citrus junos* Sieb. ex Tana.k.a.) on Mood States and Salivary Chromogranin A as an Endocrinologic Stress Marker." *Journal of Alternative and Complementary Medicine*, vol. 20, no. 6, June 2014, p. 500–506.

Nelson, Vern. "Yuzu Is Delightfully Fragrant and Unfussy." *The Oregonian*, Dec. 30, 2010. www.oregonlive.com/hg/index.ssf/2010/12/vern_nelson_hardy_citrus_yuzu.html. Accessed 08/11/18.

Ortho All About Citrus and Subtropical Fruits. Des Moines, IA: Meredith Books, 2008.

Saibante, Carola Traverso. "Yuzu Citrus from A to Z: 26 Things to Know." Finedining lovers.com. March 30, 2017. www.finedininglovers.com/stories/what-is-yuzu-citrus/. Accessed 08/06/18.

"6 Yuzu Fruit Benefits." Dr. Axe. https://draxe.com/yuzu-fruit/ Accessed 08/12/18.

Stewart, Victoria. "It's in Cocktails, Salads, Sushi and Ice Cream: The Japanese Yuzu Is Giving a Citrus Twist to Our Summer." *Evening Standard*, April 22, 2015, p. 28.

"Weird Fruit a Winner as Ugly Yuzu Proves a Flavour Hit with Aussie Chefs." *ABC (Australia Broadcasting Corporation) Regional News.* July 25, 2017.

Wong, James. "Gardens: The New Yuzu." *The Guardian*, October 18, 2015. www.theguardian.com/lifeandstyle/2015/oct/18/how-to-grow-and-eat-yuzu. Accessed 08/17/18.

Yang, H.J., et al. "Yuzu Extract Prevents Cognitive Decline and Impaired Glucose Homeostasis in β-amyloid-infused Rats." *Journal of Nutrition*, vol. 143, no. 7, July 2013, p. 1093–1099.

"Yuzu." Citrus ID, Edition 2, October 2011. http://idtools.org/id/citrus/citrusid/factsheet.php?name=Yuzu. Accessed 08/16/18.

"Yuzu ichandrin (Papeda Hybrid)." Citrus Variety Collection, University of California at Riverside, College of Agriculture and Natural Resources. www.citrusvariety.ucr.edu/citrus/yuzul.html. Accessed 08/06/18.

"Yuzu Juice—The Citrus Superfood." *Daily Mail*, November 14, 2013, p. 35.

Appendix 1. Food Safety Regarding Fruits and Berries

Food borne bacteria can not only make you sick, they can be lethal. There are a few simple easy steps you can take to help minimize your risk of food borne illness. And of course the one rule we all know is to wash our hands before and after handling fruits and berries.

- First, before consuming them or cutting them for preparation, **always** wash fresh fruits and berries in water, using a small vegetable brush if necessary to remove any surface dirt, and dry them with a paper towel.
- Always wash your hands, cutting utensils and cutting boards with soapy water after cutting fresh fruits and berries.
- Never use the same cutting utensils and cutting boards you just used to cut meat, fish, poultry, vegetables or dairy without first washing them in soapy water to avoid any cross contamination of bacteria.
- All fresh fruits and berries you cut or peel at home should either be consumed immediately, or refrigerated within two hours (at 40 degrees F or lower).
- All fresh fruits and berries that have been cut or peeled and left at room temperature for more than two hours should be discarded.
- To avoid any cross contamination, store fresh fruits and berries separate from meat, poultry, vegetables and dairy products.
- If you plan to can fresh fruits or berries, be sure to follow instructions and recipes only from current reliable sources, such as the USDA, Cooperative Extension Services, canning equipment manufacturers, and pectin manufacturers. Their recipes and instructions have been scientifically reviewed and tested to insure their safety.

- Trust only jam, jelly, conserves, fruit butters and other soft jelled fruit recipes, and pickled recipes from reliable sources like those listed above. They've made sure the recipes have the proper pectin and acidity to avoid potential botulism.

- Avoid canning instructions and recipes from unreliable sources on the internet, from old canning publications, old cookbooks, and anyone who is not a currently trained or educated expert in current food preservation science.

- Once you find a reliable recipe for canning a specific fruit or berry, be sure to follow the instructions on using the right canner (i.e., pressure canner or boiling water canner.)

- Never use the old fashioned "open kettle method" of canning; has been deemed unsafe.

- If you plan on freezing fruit, be sure your freezer is set at zero degrees Fahrenheit or lower. (*Clostridium botulinum,* the micro organism that produces a deadly toxin, does not grow or produce toxin at 0 degrees Fahrenheit.) Following freezing instructions from reliable sources will produce the best frozen products. The enzymes in fruits can cause them to brown unless pretreated as recommended for specific fruits.

- Information on safe fruit preservation methods and recipes can be found at the National Center for Home Food Preservation website at www. nchfp. uga.edu and at the USDA's National Institute of Food and Agriculture website at https://nifa.usda.gov/.

- If you have questions regarding whether a specific canning recipe is safe to use, or to have your pressure canner gauge tested, contact your state Cooperative Extension office for assistance. If you need help locating your local Cooperative Extension office, you can visit the USDA National Institute of Food and Agriculture's website https://nifa.usda.gov/land-grant-colleges-and-universities-partner-website-directory?state=All&type= Extension for a state directory.

Appendix 2. USDA Plant Hardiness Zones

In 2012 the USDA determined 13 plant hardiness zones in the United States and its territories. Each hardiness zone reflects extreme minimum winter temperatures observed from 1976–2005. If you are selecting a plant for your home garden, it's important to make sure the plant is hardy in your USDA zone so that it will survive your cold winter temperatures. An alternative is of course, to grow a plant that is not hardy for your zone in a temperature controlled greenhouse. If you don't know what USDA hardiness zone you live in, you can go the USDA Plant Hardiness map website https://planthardiness.ars.usda.gov. Once there, you can type in your zip code to find out your plant hardiness zone.

It should be noted that the USDA hardiness zones reflect the range of coldest temperatures, but do not reflect the heat or humidity of a zone. Many subtropical and tropical plants need a specific amount of sunlight, heat and or humidity both to fruit, and to have the fruit ripen. To determine if your zone has the necessary heat and humidity for a specific fruit plant to grow, you can contact your local Cooperative Extension service for assistance. If you need assistance locating your local Cooperative Extension office, you can visit the USDA National Institute of Food and Agriculture's website https://nifa.usda.gov/land-grant-colleges-and-universities-partner-website-directory?state=All&type=Extension for a state directory.

USDA Plant Hardiness Zones (including half zones)
Range of Minimum Average Annual Winter Temperatures

Zone	Temperature (Fahrenheit)	Temperature (Celsius)
1a	-60 to -55	-51.1 to -48.3
1b	-55 to -50	-48.3 to -45.6
2a	-50 to -45	-46.6 to -42.8
2b	-45 to -40	-42.8 to -40.0
3a	-40 to -35	-40.0 to -37.2
3b	-35 to -30	-37.2 to -34.4
4a	-30 to -25	-34.4 to -31.7
4b	-25 to -20	-31.7 to -28.9
5a	-20 to -15	-28.9 to -26.1
5b	-15 to -10	-26.1 to -23.3
6a	-10 to -5	-23.3 to -20.6
6b	-5 to 0	-20.6 to -17.8
7a	0 to 5	-17.8 to -15.0
7b	5 to 10	-15.0 to -12.2
8a	10 to 15	-12.2 to -9.4
8b	15 to 20	-9.4 to -6.7
9a	20 to 25	-6.7 to -3.9
9b	25 to 30	-3.9 to -1.1
10a	30 to 35	-1.1 to 1.7
10b	35 to 40	1.7 to 4.4
11a	40 to 45	4.4 to 7.2
11b	45 to 50	7.2 to 10.0
12a	50 to 55	10.0 to 12.8
12b	55 to 60	12.8 to 15.6
13a	60 to 65	15.6 to 18.3
13b	65 to 70	18.3 to 21.1

Appendix 3. Chilling Hours

A number of fruit trees require what are known as "chilling hours." Chilling hours are hours when the temperature remains between 32 degrees and 45 degrees Fahrenheit (or 0 degrees to 7.2 degrees Celsius). Chilling hours are important because they help regulate plant growth. During the winter many fruit bearing plants go dormant to protect themselves from freezing cold weather. During dormancy plants produce hormones (growth inhibitors) that prevent the plant from growing, despite extreme fluctuations in the temperature that may occur. Through thousands of years of plant evolution, chilling hours tell the plant the minimum amount of time it needs to remain dormant. Once the minimum chill hours have been met, and the temperature warms up again, the plant breaks dormancy and resumes growth.

If a plant doesn't get the necessary chill hours, it interrupts its natural process and creates problems. Think of it like a human who gets less than two hours of sleep a night instead of eight. Due to the lack of sleep, that person will not be able to function normally and the lack of sleep will result in his having numerous problems during the day. If a fruit plant doesn't receive the minimum number of chill hours it requires, it will also not function as it should. It may fail to develop blossoms, the blossoms and new leaves may appear later than normal. The plant may develop little or no fruit, and any fruit produced may be smaller than normal or malformed. You can find methods for determining your local chill hours on the internet. But it takes some calculations and temperature monitoring during the winter months. An easier, more reliable way to find out the chill hours in your area is to contact your local agriculture department or local cooperative extension office.

If you need assistance locating your local Cooperative Extension office, you can visit the USDA National Institute of Food and Agriculture's website https://nifa.usda.gov/land-grant-colleges-and-universities-partner-website-directory?state=All&type=Extension for a state directory.

Below are the chilling hours required for Asian fruits listed in this book.

Fruit	Minimum Chilling Hours*
Asian persimmons	100–200
Asian pears	350–450
Fig	100
Kiwi	50–800
Lychee	100–200
Nanking cherry	200–300
Pomegranate	100–150
Quince	300

*When a range is given, the minimum chilling hours required varies by cultivar.

Appendix 4. Fertilizers

The first time you shop for synthetic fertilizers can be confusing. You'll see a variety of fertilizer products on store shelves, some with labels showing the letters N-P-K on them followed by three varying numbers, some with a variety of other letters like Zn and Fe. The N on labels represents nitrogen. Plants use nitrogen to build new cells, and in turn to grow new shoots and leaves. Excessive nitrogen can increase a plant's susceptibility to pests and diseases. The P on labels represents phosphorus, which is needed by plants to help them grow strong roots. The K on labels represents potassium, which helps the plant regulate photosynthesis.

Other letters that may appear on labels include: Zn for zinc often appears in foliar fertilizers. Fe for iron, also frequently appears in foliar fertilizers. B represents boron. Ca represents calcium. Mg represents magnesium. S represents sulfur. All of these elements play a role in the growth of fruit bearing plants.

The numbers that follow the letters show the amount of the substance in the bag. For example, if a bag label reads N-P-K 30-10-30, it means the bag contains 30 percent nitrogen, 10 percent phosphorus, and 30 percent potassium. So if you purchase a 100-pound bag of fertilizer labeled N-P-K 30-15-15, it contains 30 pounds of nitrogen, 15 pounds of phosphorus and 15 pounds of potassium. The remaining 40 pounds in the bag is filler.

Nitrogen, phosphorus and potassium can also be obtained from organic fertilizers, such as compost, manure, bat guano, cottonseed meal, etc. The difference between organic and synthetic fertilizers is that synthetic fertilizers are released more rapidly and are therefore fast acting, and usually less expensive. But they also have the potential for environmental contamination from runoff if they are over applied.

Fertilizer should be applied to correct deficiencies in your soil that affect plant growth. Soil testing kits that analyze your soil can be purchased at garden centers, or better yet, you can have your soil tested in a soil lab. Contact your local Cooperative Extension Office for information on soil testing labs in your area. And if you do need to apply fertilizer to your soil, your local Cooperative Extension Office can provide you with information on alternative organic fertilizers, the amount of fertilizer to apply as well as when and how often to apply the fertilizer.

If you need assistance locating your local Cooperative Extension office, you can visit the USDA National Institute of Food and Agriculture's website https://nifa.usda.gov/land-grant-colleges-and-universities-partner-website-directory?state=All&type=Extension for a state directory.

Glossary

Air layering A propagation method where the soil or rooting medium is bound to the plant. A slanting cut is made into a one to two year old branch. The wound is frequently dusted with a rooting hormone and a small amount of sphagnum moss is packed into the wound to keep it from healing over. Moist sphagnum moss is wrapped around the wound until it develops roots.

Apex The top or tip of an object.

Apomictic seed Clones of the parent plant produced through asexual reproduction. So plants grown from apomictic seeds are exact clones of the parent plant.

Budding A propagation method where a bud is taken from one plant and grown on another. This is also known as bud grafting.

Calyx Usually green, the calyx consists of sepals that surround and enclose the flower bud.

Chilling hours Hours when the temperature remains between 32 and 45 degrees Fahrenheit (or 0 to 7.2 degrees Celsius).

Cleft grafting This is a method of grafting where the stub on a branch on the stock is split open and the scion is inserted into it.

Cultivar A cultivar is a plant variety that's been developed through selective breeding.

Deciduous Trees and other plants that shed their leaves in the fall.

Espalier Training a tree or bush to grow flat against a wall through pruning and other methods.

Germination The process where a new plant is grown from a seed or other spore.

Grafting A propagation method where the scion (the upper part) of a plant is grown on the rootstock of another plant.

Hilum The line or scar on a seed that indicates where it was attached to the embryo sac.

Marcottage *see* **Air layering**

Propagation The process of creating new plants.

Recalcitrant seeds Seeds that are short lived and do not survive drying.

Scion A plant twig or branch shoot used for grafting to a plant stock.

Seed scarification The weakening or modification of the seed coat to improve seed germination rates. Scarification is often done thermally (soaking the seeds in hot water), mechanically (nicking or opening the seed coat), or chemically (dipped into sulfuric acid, acetone or other organic solvent).

Seed stratification The process of breaking a seed's dormancy to promote germination. In this process the environmental conditions the seed requires to break dormancy naturally is mimicked, thus the stratification process varies by seed. Some seeds need a cold period to break dormancy, while others may need a warm moist environment to break dormancy. In cold seed stratification, a seed is kept moist and subjected to cold (usually in a refrigerator) to a certain amount of time.

Soil pH This is a measure of the soil's acidity or alkalinity. A soil pH of 7 is considered neutral. And the lower the number reflects greater acidity. The higher the number reflects greater alkalinity. Plants generally grow best in neutral or slightly acidic soil.

Stock Also known as rootstock, a plant onto which a bud or shoot is grafted.

Testa The protective covering on a seed.

Vegetative propagation This refers to asexual reproduction of a plant. Basically a piece of a plant is taken to produce more plants. Examples of vegetative propagation include grafting, cuttings, layering, suckering, etc.

Veneer grafting In this type of grafting, an angled wedge cut is made of the side of the stock, instead of the top. A similar angled cut is made to the scion so it can be inserted into the stock, then wrapped with grafting tape to hold it in place.

Index